PENGUI

WORKING

William Rukeyser is the managing editor of *Fortune*. Contributing editors include Jeremy Main, William Bowen, Edward Meadows, Robert Lubar, Charles G. Burck, Gene Bylinsky, and Irwin Ross.

Introduction by

WILLIAM S. RUKEYSER

PENGUIN BOOKS

Penguin Books Ltd, Harmondsworth,
Middlesex, England
Penguin Books, 40 West 23rd Street,
New York, New York 10010, U.S.A.
Penguin Books Australia Ltd, Ringwood,
Victoria, Australia
Penguin Books Canada Limited, 2801 John Street,
Markham, Ontario, Canada L3R 1B4
Penguin Books (N.Z.) Ltd, 182–190 Wairau Road,
Auckland 10, New Zealand

First published in the United States of America by
The Viking Press 1982
Published in Penguin Books 1984

LIBRARY OF CONGRESS CATALOGING IN PUBLICATION DATA
Main entry under title:
Working smarter.
Reprint. Originally published: New York: Viking Press, 1982.
Bibliography: p.
Includes index.
1. Industrial productivity—United States—Addresses, essays, lectures.
2. Technological innovations—United States—Addresses, essays, lectures.
3. Competition—United States—Addresses, essays, lectures. 4. Organizational effectiveness—Addresses, essays, lectures. I. Fortune.
[HC110.I52W67 1984] 338′.06′0973 83-25696
ISBN 0 14 00.6894 5

Printed in the United States of America by
R. R. Donnelley & Sons Company, Harrisonburg, Virginia
Set in Linotron Times Roman

All the articles in this book appeared originally in *Fortune* magazine.

The charts that appear on pages 4, 5 and 15, 94, and 143 and 146 were redrawn from originals by Joe Argenziano, Luis Perelman, Arnold Parios, and William Reduto, respectively.

CONTENTS

INTRODUCTION vii

1 THE PROSPECTS FOR PRODUCTIVITY 1
by William Bowen

2 THE NEXT INDUSTRIAL REVOLUTION 19
by Gene Bylinsky

3 SUCCESS AT THREE COMPANIES 33
by Edward Meadows

4 REDISCOVERING THE FACTORY 44
by Robert Lubar

5 THE QUALITY OF WORKLIFE 57
by Charles G. Burck

6 WESTINGHOUSE'S CULTURAL REVOLUTION 70
by Jeremy Main

7 BATTLING YOUR OWN BUREAUCRACY 81
by Jeremy Main

8 CAN DETROIT CATCH UP? 92
by Charles G. Burck

9 **WHEN WORKERS MANAGE THEMSELVES** 106
by Charles G. Burck

10 **WHAT'S IN IT FOR THE UNIONS** 118
by Charles G. Burck

11 **A NEW SPIRIT IN ST. LOUIS** 128
by Irwin Ross

12 **WHY GOVERNMENT WORKS DUMB** 140
by Jeremy Main

13 **THE BATTLE FOR QUALITY** 152
by Jeremy Main

14 **SERVICE WITHOUT A SNARL** 164
by Jeremy Main

15 **SHARPENING THE COMPETITIVE EDGE** 177
by William Bowen

BIBLIOGRAPHY 195

APPENDIX 199

INDEX 203

INTRODUCTION

The U.S. achieved economic preeminence by producing more goods and better goods with less labor—in short, by working smarter—than anyone else. The U.S. economy is still the most productive in the world, but success can no longer be taken for granted. In the years since World War II, Japan, Italy, West Germany, France, Canada, and even poor Britain have all outstripped the U.S. in productivity gains. As the 1980s began, U.S. industry could no longer ignore the growing evidence that it was being beaten at its own game. The Japanese had invaded and then conquered one market after another and set new world standards for quality. Their TV sets, stereos, automobiles, cameras, and many other exports achieved levels of quality never seen before in mass-produced consumer goods.

Productivity is the essential ingredient of a rising standard of living. People can produce more goods and services for themselves only by working longer hours or by improving their productivity. And quality is almost inseparable from productivity. When quality is low, output is reduced to the extent that faulty items have to be discarded, repaired, or replaced. Conversely, one of the cheapest ways to raise productivity is to raise quality—a cut in scrapped parts adds automatically to output.

In 1980 the U.S. woke up to a productivity-quality emergency.

Productivity had actually declined in the two previous years. Huge losses in the automobile industry, the possible collapse of Chrysler, and the surrender of 27 percent of the domestic auto market to Japanese manufacturers were all chilling evidence of what many other industries faced if they failed to make quality and productivity improvements their major goals.

In a preview of the economy of the 1980s *Fortune* magazine singled out productivity as a basic problem for the U.S. (see Chapter 1). As businessmen and economists became more concerned, *Fortune* returned to the subject in a 1981 series entitled "Working Smarter." These and related articles are the basis of this book.

The *Fortune* editors, writers, and reporters who prepared the articles discovered that the more they searched, the more they found that many answers to quality and productivity problems had been ready to be picked like low-hanging fruit for decades. Way back in 1942 *Fortune* published an article entitled "65 Billion Man-Hours" by a consultant named Allan H. Mogensen, who suggested approaches that are only today gaining wide acceptance. Mogensen taught industrial engineering at the University of Rochester in the 1920s in the tradition of Frederick W. Taylor, the man who used time-and-motion studies to reduce jobs to the simplest and most efficient functions. But Mogensen became aware that another element was needed to make a success of "work simplification." He came to the conclusion that "the person doing the job knows far better than anyone else the best way of doing that job and therefore is the one person best fitted to improve it." He was a pioneer in the now popular concept of "participative management" (see Chapters 5 and 6). Mogensen also said people had to "work smarter, not harder"—a phrase that explained so well what we need today that it provided the title of this volume.

Mogensen found out in the 1930s that it would be possible to increase the output of a work force by as much as 50 percent. At first *Fortune*'s editors didn't believe these claims. Before printing his article they sent reporters out to investigate companies that were using his methods. The claims were confirmed, and today, at 80, Mogensen is still giving the annual productivity seminars he started in 1937 and still piloting his own plane.

Enthusiasm in these matters seems to carry with it the gift of permanent youth: the prophets of quality improvement are Mogensen's contemporaries and remain just as active. W. Edwards Deming, 81, and Joseph M. Juran, 76, developed their basic ideas about improving quality decades ago, but their fellow Americans didn't listen closely enough. The Japanese did, though, and Deming and Juran became heroes on the other side of the Pacific. Only now is the significance of their work getting full recognition in the U.S. (see Chapters 13 and 14).

Of course, theories about productivity and quality were not carved in stone years ago for all time. They keep evolving, as Chapter 5 relates. Until recently, the better use of better machinery was considered the key to improving productivity, and that approach still has merit. Chapter 2 reports how much promise of productivity lies in the world of CAD/CAM (Computer Aided Design/Computer Aided Manufacturing). Chapter 3 tells how three companies have raised output in relatively traditional fashion, by improving the production line. Chapter 4 reports that the factory manager, after years in eclipse, is recognized once again as a vital executive.

But looking for all the solutions to productivity problems on the production line is too restrictive. Investment in plant and equipment per worker—a standard for predicting productivity by which the U.S. lags behind West Germany and Japan—is losing its significance because factories account for less of the economy's total output. Many manufacturing companies pay out more in white-collar salaries than in blue-collar wages nowadays, and the office offers at least as much opportunity as the factory for improving output. Chapters 6 and 7 examine how two companies, Westinghouse and Intel, are promoting productivity in the office.

The fears and hopes about American productivity and quality seem to be epitomized by the troubles of Detroit. Chapter 8 describes how far behind Japan Detroit has fallen—and what an opportunity the U.S. auto industry has to pull off a historic turnaround. *Fortune* might have discussed other industries and other companies had space permitted. For instance, since its establishment in 1939, Hewlett-Packard, one of the most successful exploiters of modern technology, has practiced the participative style. Procter & Gamble turned to work simplification in 1946 and developed what it calls a "deliberate methods change program." Combined with other cost savings, the program cut the company's expenses in fiscal 1981 by $803 million—more than the year's net earnings. Lockheed, General Motors, and a few other companies began setting up "quality circles" early in the 1970s. (For a discussion of quality circles, see Chapter 9.) As the 1980s got under way, a thousand more corporations rushed to adopt quality circles.

With change comes a danger of voguishness. Businesses misapply good new ideas or apply bad ideas and then withdraw in disappointment. Some of the ideas advanced in this book will in the end doubtless prove unworkable in the U.S. or in certain companies, but many others seem sure to work well.

The urge to improve productivity and quality—so strong because in some cases it involves literally the survival of corporations—is changing the traditional hostility between business and labor. The two sides are beginning to see they share goals. How workers and unions are taking to the new concepts is discussed in Chapters 9 and 10, and a successful plan to raise productivity and cut jurisdictional squabbles in the construction industry in St. Louis is described in Chapter 11. Government also faces the need to improve its performance. Indeed, the gains in output per worker that have been achieved in many companies would, if applied in Washington, make a sizable dent in the federal deficit. What can be done in government—and how much more difficult it is to do there than in the private sector—is discussed in Chapter 12.

This book concludes with an article from a special Time Inc. editorial project, "American Renewal," in which, early in 1981, *Fortune* assistant managing editor William Bowen laid out a program for helping cure America's productivity sickness. Some of the medicine he prescribed has since been administered by the Reagan Administration, but much remains to be done.

Articles are reprinted in this volume as they first appeared in *Fortune,* with only minor changes. We have resisted the temptation to imbue the chapters with hindsight. However, important related events or changes that have occurred since the articles were first printed are footnoted.

The book is the work of many editors, writers, and reporters at *Fortune,* among whom I want to pay special tribute to executive editor Allan T. Demaree, who conceived and planned the "Working Smarter" series, and to the principal authors of that series, Charles G. Burck and Jeremy Main. Main, a member of *Fortune*'s board of editors, also took charge of adapting the articles for this book. Associate editor Anna Cifelli scoured the manuscript for accuracy and consistency. The book is also the work of many, many managers, workers, clerks, academics, union leaders, politicians, and government administrators, who spent hundreds of hours talking to us about productivity and quality. We are most grateful to them, but we did not feel we were imposing on them. Most were unselfishly eager to share with us their enthusiasm for what they were learning about how Americans can work smarter.

William S. Rukeyser
Managing editor, *Fortune*
New York, N.Y.
April 1982

WORKING SMARTER

I

THE PROSPECTS FOR PRODUCTIVITY

WILLIAM BOWEN

"The alarm bells are finally beginning to ring," said a witness before the Joint Economic Committee of Congress in 1979. He was C. Jackson Grayson, Jr., chairman of the American Productivity Center in Houston, and he was testifying about the deterioration in the growth of American productivity—a "virtual collapse," as one study put it. He certainly proved to be right about alarm bells; they seem to be ringing all the time. A productivity buff could keep busy just going to seminars, symposiums, conferences, and lectures on the subject. Almost every week seems to bring an announcement of another panel being established somewhere to study our ailing productivity and—presumably—to propose things to do about it.

Unfortunately, all the ado about productivity is thoroughly justified. The falloff in productivity growth, says Burton Malkiel, chairman of the economics department at Princeton, is "the most basic sickness of the U.S. economy." Productivity performance is perhaps the best single indicator of an economy's vitality. Rising productivity, after all, is where gains in standards of living come from. Economist Roger Brinner of Data Resources, Inc., makes the often overlooked point that the reason for the lack of growth in real income per worker in recent years is not inflation, as most people believe, but the lack of growth in productivity. (Inflation, to be sure, helps retard productivity.) Poor productivity growth, moreover, has weakened the ability of the U.S. to compete in world markets, and the implications of continued enfeeblement are exceedingly unpleasant.

The deterioration in American productivity growth since the middle 1960s reversed a trend that had been running since the early years of the nation's history. In a long perspective, smoothing out the cyclical fluctuations that productivity is subject to, it is evident that the rate of productivity growth in the U.S. moved in a generally rising trajectory for nearly two hundred years before turning downward in the late 1960s. Data put together by George Washington University's John Kendrick, a leading authority on productivity, indicate that output per worker grew at a rate of 0.5 percent a year over the 1800–55 span, 1.1 percent in 1855–90, 2.0 percent over the next three decades, and 2.4 percent in 1919–48, despite the downdrag from the Great Depression. From then until the middle 1960s, the trend rate was 3.2 percent. (These figures, and all those having to do with productivity in this chapter, apply only to the private business sector, which excludes government, nonprofit organizations, and households. Productivity growth rates for the entire economy, including government, would be somewhat lower.)

In 1966 the Council of Economic Advisers formally adopted 3.2 percent as the supposedly noninflationary "guidepost" for union wage negotiations. It soon became obsolete. Following a premonitory stumble in 1967, productivity grew only 0.2 percent in 1969 and then only 0.7 percent the following year. There was some recovery after that, but productivity growth never did get back to the 1947–66 trend line. Instead, a new and still weaker trend set in after the great leap in OPEC prices in 1973. The rate for 1973–78 was only 1.1 percent.

In the 1980s things will get somewhat better. The projections of Alan Greenspan, who collaborated with *Fortune* on a series about the U.S. economy in the decade ahead, call for a productivity growth rate of 2.1 percent. This moderate recovery will at least relieve worries that stagnation might be just around the corner.

But the prospective recovery, welcome as it will be, will bring us back only to the slowed-down 2.1 percent trend rate of 1966–73—not even close to the old 3.2 percent. Even this blue-gray outlook is predicated on some fairly optimistic assumptions. The projections for the 1980s involve certain assumptions about oil and about the competence of the national government. It was assumed that we won't run into severely damaging disruptions of our oil supply and that oil prices won't rise much faster than the general price level, and it was assumed that, as Greenspan put it, "We won't shoot ourselves in the foot" in our policies and behavior as a nation. These assumptions are not mush, but no sensible person would bet his whole bundle on them. So even a comeback to 2.1 percent in productivity can't be counted as a sure thing.

The slowdown in productivity growth has been all the more troubling because economists have found it so hard to explain—or explain convincingly. In the words of John P. White, deputy director of the Office of Management and Budget, "There is no single or even dominant explanation, and there is no general agreement about the quantitative effects of various factors." There is widespread agreement that the slowdown has gone through two periods, with 1973 the dividing year, but economists disagree as to which period is the more mysterious. Edward F. Denison, a renowned expert on economic growth, thinks he's got the first phase pretty well explained, but he finds the second one "a mystery." In contrast, J. R. Norsworthy, head of productivity research in the Bureau of Labor Statistics, can account to his satisfaction for most of the slowdown since 1973, but he is "puzzled" by the earlier period.

Though its behavior over time can baffle even experts, productivity is a seemingly straightforward concept: real output per worker-hour. (That's the American version; in most other countries that keep data, it's real output per worker.) Economists sometimes seem overly concerned about the possibility that laymen will think improvements in output per worker-hour come entirely from the labor side—because workers are working harder or better. In 1979, for example, a panel assembled by the National Academy of Sciences solemnly recommended that the

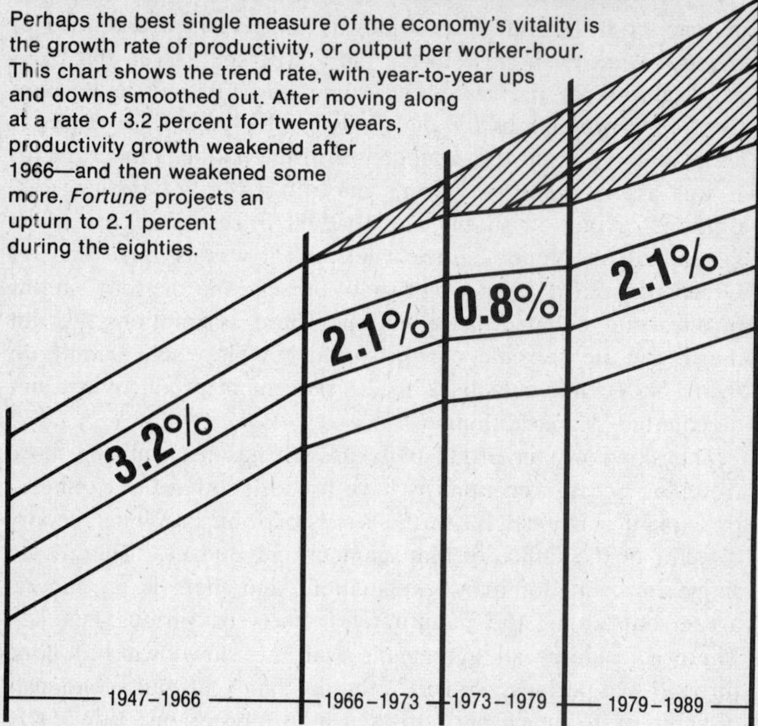

Bureau of Labor Statistics "give more prominence to cautionary statements" to keep people from supposing that changes in productivity result solely from "the changing skill and effort of the work force." It is hard to believe that any red-blooded American businessman thinks anything of the sort, but just in case, let it be firmly said that improvements in productivity come largely from nonlabor factors, notably technology and tangible capital.

Over any short run, changes in measured productivity are dominated by cyclical influences. Measured year to year, as the chart shows, the rate of change in productivity is a jumpy indicator. Over a long run, the cyclical bouncing dwindles in importance, but it does not fade out altogether, because the economy's short-run fluctuations influence basic factors of growth, such as capital formation.

HELP FROM DEMOGRAPHICS

Productivity growth over a span of years is affected by a great many interacting influences, but most of the change from one

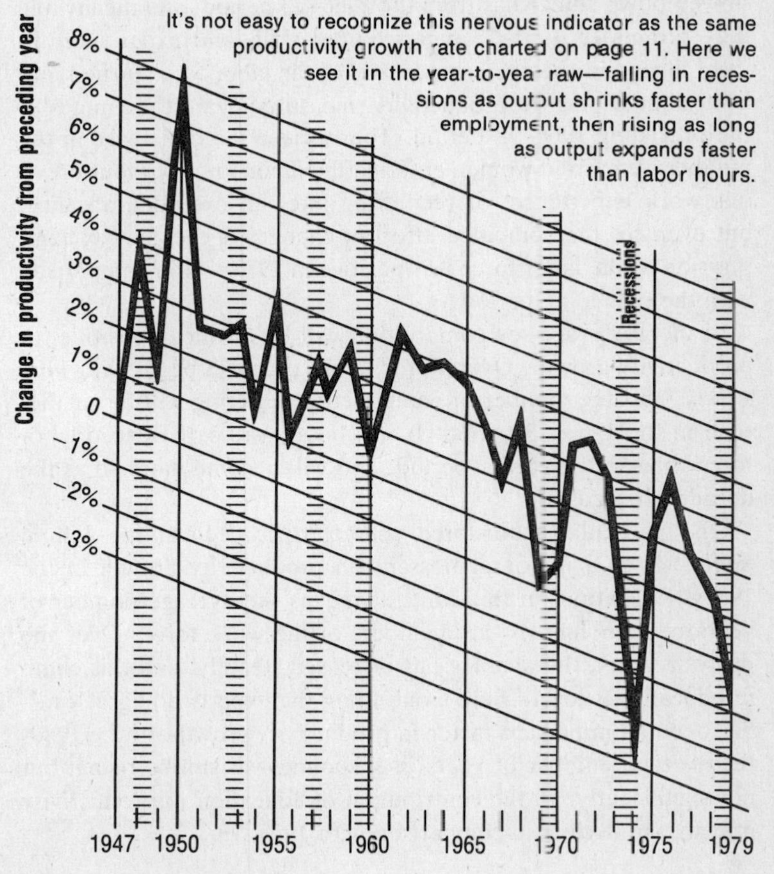

period to another can be explained in terms of half a dozen factors. One of these is changes in characteristics of the labor force—levels of skill, experience, education. A major influence on productivity in the 1966–73 period was the surge in the numbers of young people entering the labor force—the baby-boom kids of the 1940s and 1950s arriving at working age. This influx is counted as a large minus for productivity because the disproportionately large numbers of young people lowered the average levels of skill and experience in the work force.

In the next period, 1973–78, the age composition of the labor force was no longer a depressant—the influx of young people slowed down somewhat from the 1966–73 period, and meanwhile the entrants of 1966–73 had acquired skills and experience. In fact, change in the age composition of the labor force turned into a substantial plus for productivity, measured against the minus of the preceding 1966–73 period. There was a large increase in the number of married women entering the labor force without previous work experience, or reentering after an extended absence, but even so, the combined effect of changes in the age-sex composition of the labor force was positive in 1973–78, in comparison with the preceding period.

In the 1980s age-sex composition will be distinctly positive for productivity growth. Overall, the labor force will be growing very slowly, and the number of young people entering will be smaller than in the 1966–73 period. It will be a more experienced labor force than in the earlier period, and will become more so as the decade proceeds.

One particular labor-force characteristic—education—will in effect act as a major depressant on productivity growth in the 1980s. Education, in this context, means the average number of years of schooling of the members of the work force. Over the postwar years, the average has increased steadily, and this climb in educational level—here's one thing the experts do agree on—has been an important factor in productivity growth. In the 1980s the average number of years of schooling will still be rising, but not significantly, so the contribution of education to productivity growth will decline in comparison with 1973–78.

An important determinant of productivity is the stock of invested capital per worker—the capital-labor ratio. There is fairly general agreement that the rate of growth of this ratio did not change much from 1947–66 to 1966–73, and therefore did not account for any of the deterioration in productivity growth between those two periods. But the 1973–78 period is quite another story—a slackening in the growth of the capital-labor ratio was the principal depressant, accounting for well over half of the fall-off in productivity growth.

The decline was the combined result of a large increase in employment and an extraordinarily small increase in capital formation. Over the 1947–73 span as a whole, business fixed investment grew at a rate of 3.9 percent a year. In 1973–78 it grew at a feeble 1.4 percent, less than the growth rate of employment during that span. No wonder productivity growth was weak.

The sluggishness in capital spending can be largely explained as a consequence of the resurgence of inflation after the 1974–75 recession. Inflation discourages capital investment. Among other things, it pushes up interest rates, worsens perceived risk-reward ratios, and shortens time horizons. (Management consultant James Skidmore, head of Science Management Corporation, recently observed that in many companies these days, "the five-year plan has become the two-year plan.")

In the 1980s, according to the *Fortune*-Greenspan projections, the capital-labor ratio will recover strongly, largely as a result of the assumed slowing of inflation. Measures to reduce the tax load on business will also help. The growth in capital investment will be by far the strongest productivity-buoying factor during the decade.

This prospect, however, will be dimmed if oil prices rise considerably faster than prices generally, lifting the real prices of other forms of energy as well. That would serve to depress capital investment. *Fortune*'s basic projection of 3.4 percent growth in real Gross National Product (GNP) and 2.1 percent growth in productivity was predicated on no growth in the real price of oil after the already signaled rise in 1980. If, for example, oil prices were to increase 5 percent a year in real terms, the GNP would grow at a rate of only 3.1 percent, and productivity at only 1.9 percent.*

THE FALLACY ABOUT SERVICES

Another significant factor in changes in measured productivity is what economists call "intersectoral shifts." Some industries or sectors of the economy have much higher productivity than others. When there are relative increases in the numbers of workers in sectors of higher (or lower) productivity, this can affect the overall trend rate aside from any changes in productivity *within* the sectors.

The most important intersectoral shift over the years was the movement of workers from farms to factories and other workplaces where productivity was higher. The migration out of farming, however, pretty much ended in the first half of the 1960s, and this fade-out of an important source of productivity growth accounts for a significant portion of the drop in the productivity growth rate for 1966–73.

It is widely believed that the shift to services—that is, relative growth in employment in services, as compared with manufacturing—has been a major depressant for productivity growth, and indeed some economists say so. For example, M.I.T. professor Lester Thurow puts the shift to services first among the causes of the productivity slowdown. Since the average level of productivity is lower in services than in manufacturing, or in the economy generally, it certainly seems that a relative shift of employment to services *should* tend to retard productivity growth. The fallacy lies in thinking of services as a single sector with a single level of productivity, when in fact there are diverse service sectors with very different levels of productivity. Much of the growth in service employment since the middle 1960s has come in service sectors with relatively high levels of productivity—communications, financial services, and air transportation, to name a few. In the words of Jerome A. Mark, assistant commissioner of the Bureau

*After they took that big 130 percent jump early in 1980, oil prices did stabilize, and then declined in 1982.

of Labor Statistics, "The shift to services can be viewed only as a very minor source of the slowdown."

THE COST OF REGULATIONS

A far from minor influence on productivity growth since 1966 has been the increase in burdens imposed on business by federal regulation, which started to escalate briskly in the late 1960s. These burdens are expected to level off around the middle of the 1980s, so the regulation factor will be a plus for productivity growth—it will still be a major depressant on the *level* of productivity in the U.S., but the costs of complying will not increase as rapidly as in 1973–78, so the effect on the rate of productivity *growth* in the 1980s, as compared with 1973–78, will be positive.

However, a vigorous challenge to this conventional view comes from economist Murray Weidenbaum, head of the Center for the Study of American Business at Washington University. As he sees it, "The worst is ahead, rather than behind us." A close student of regulation and the author of a book on the subject (*The Future of Business Regulation*), Weidenbaum sees potential for huge increases in costs from some of the existing statutes and rulings, including the Toxic Substances Control Act and the Occupational Safety and Health Administration's carcinogen standards.

Support for Weidenbaum's dissenting view can be heard at the American Productivity Center. Michael Lang, a lawyer who serves as the center's lobbyist in Washington, predicts that "OSHA rules are going to get worse and worse and worse as far as expense is concerned." The root trouble, he says, is that the original OSHA enactment, like some environmental legislation, did not set any cost standards or limits. But there's a new awareness in Congress these days that regulatory costs are excessive, and some movement in the right direction may come of that.*

Yet another important influence on productivity is the growth rate of the economy—growth, it seems, favors growth. Strong

*President Reagan appointed Murray Weidenbaum chairman of his Council of Economic Advisers when he took office, and the Reagan Administration has moved to reduce the regulatory burden on business.

productivity growth lifts GNP growth, and strong GNP growth feeds back on productivity. Vigorous growth of the economy fosters confidence, opens up opportunities for investment, generates profits to finance investment, stimulates innovation and the diffusion of new technology, and makes possible increased economies of scale; stagnation has all the contrary effects. Call this the "buoyancy" factor. GNP grew at a slower rate in 1966–73 than it did over the 1947–66 span, and growth was even more sluggish in 1973–78, so the buoyancy factor was a minus in both periods. But in the 1980s, on the *Fortune*-Greenspan projections, GNP growth will speed up, so the buoyancy factor becomes a plus.

Together these six causal factors—age-sex composition, education, capital-labor ratio, intersectional shifts, regulation, buoyancy—go a long way toward explaining both the weakness in productivity since the middle 1960s and the prospective pickup in the 1980s. But for both 1966–73 and 1973–78, the combined effects of the six factors fall short of accounting for the entire decline in productivity growth. There is an unexplained residual of about 0.3 percent for each period. For the 1980s, however, there is no residual—the six causal factors fully account for the 1 percent recovery in the rate of productivity growth. This does not mean that the negative factors in the 1973–78 residual will disappear in the 1980s—it means that, on balance, they will stop making things worse, and so exert no further downward push on productivity growth. (The arithmetic of all this—both the two downturns in the trend rate and the partial recovery in the 1980s—is set forth in summary form in the accompanying table.)

Looking at the two periods of falloff together, we have a combined residual of 0.6 percent. Give or take a tenth of a percentage point, this figure approximately defines the area within which there is some "mystery" about the weakening of American productivity growth. The unexplained residual, of course, is mysterious only because it resists explanation within the analytical frameworks that economists use to account for changes in productivity. Outside these frameworks, it is possible to identify a number of influences that may have had significant adverse effects on productivity.

PUSHES AND PULLS ON PRODUCTIVITY

Most of the changes in the trend rate of productivity growth over the past thirty years can be explained in terms of the six factors listed in this table. The figures in the columns are tenths of a percentage point. They indicate how much particular factors contributed to the *change* in the trend *rate* of productivity growth from one period to another. An entry of 0.0 means that the factor's contribution was the same in both periods. A lessening of an upward pull on the productivity growth rate from one period to another is the equivalent of a downward push (that's what happens to education in the final column). And a lessening of a downward push is the equivalent of an upward pull (that's what happens to regulatory costs in the final column).

	Change in trend rate of productivity growth from:		
	1947–66 to 1966–73	1966–73 to 1973–78	1973–78 to eighties
FACTOR OF CHANGE	**Down 1.1**	**Down 1.0**	**Up 1.0**
Capital-labor ratio	0.0	−0.6	0.7
Age-sex composition of labor force	−0.3	0.2	0.3
Education of labor force	0.0	0.0	−0.3
Shift out of farming	−0.3	0.0	0.0
Regulatory costs	−0.1	−0.2	0.2
Buoyancy of economy	−0.1	−0.1	0.1
SUBTOTAL	−0.8	−0.7	1.0
Residual (miscellaneous downers)	−0.3	−0.3	0.0
TOTAL	−1.1	−1.0	1.0

Quite a few people think that the weakening of productivity growth is partly a result of deterioration in attitudes toward work. Those who think so include many plant managers and economists of such divergent political viewpoints as Paul Samuelson, of textbook fame, and Herbert Stein, chairman of the Council of Economic Advisers under Presidents Nixon and Ford. Samuelson counts " a weakening of the hungriness motives" as a major reason for the slowing of economic growth in the industrial world in the 1970s. Stein talks about the effects of Me Generation attitudes. Harvard sociologist David Riesman thinks the spread of the counterculture into the labor force has damaged productivity growth by contributing to an erosion of workplace discipline and a decline in care and attention.

Stagflation itself may have some productivity-damaging effects. Certainly the combination of severe inflation, the graduated income tax, and barely visible growth in real income per worker has damaged the morale of a great many Americans in the last few years, and this may have affected performance at work to some extent. One businessman who thinks this has begun to happen is Jerome Hardy, president of Dreyfus Corporation. At a meeting with members and staffers of the Joint Economic Committee in Washington last spring, he said: "People . . . are beginning to lose faith in the whole structure, in my opinion. And I would try my damnedest to find ways to reestablish that faith, because it is exhibiting itself in a cynicism about productivity insofar as commitment to work is concerned."

A PACKAGE OF DOWNERS

While there are differences of opinion about whether changes in the attitudes of workers have damaged productivity, there is no doubt at all that crime has done so—more precisely, the increased incidence of crime. Pilferage and shoplifting pare output. Vandalism adds to costs. In efforts to protect property and employees, many companies hire extra employees to serve as guards and watchmen. That means additional input without additional output—in other words, impaired productivity.

Another likely contributor to that unexplained residual is inflationary damage to economic efficiency. Price has an important signaling function in economic life. With rapid inflation, the efficiency of this signaling function is seriously impaired. Markets become less efficient, and that entails some drag on productivity.

Some of the subtler effects of burdensome regulation, not counted in compliance costs, may also have had a negative influence. Regulatory matters absorb a significant portion of the time of top management, possibly diverting attention from such other matters as expansion of output, reduction of costs, and enhancement of productivity. More serious, perhaps, is the regulatory tar in the gears of decision-making—the continual need to take account of large numbers of complex regulations. One observer remarked that business was being "regulated into slow motion."

Murray Weidenbaum argues that affirmative action may be doing some damage to productivity. To the extent that it does no more than ensure equal opportunity, affirmative action should be good for productivity in the long run, making it possible for more people to realize their potential and thus enlarging the pool of people competing for higher-level jobs. But Weidenbaum thinks that reverse discrimination, especially where it involves a lowering of standards, can impair productivity by stirring resentment and impelling other workers to notch down *their* standards. And of course affirmative action in all its ramifications consumes a considerable amount of management time.

The combined depressant effect of this package of suspected productivity downers should certainly get no worse in the 1980s. The slowing of inflation that Greenspan sees in the cards will be a plus in more ways than one. The prospective reduction in the proportion of teenagers and young adults in the population may mean less crime. With the growing awareness in Washington that regulatory burdens on business are too heavy, there may be some leveling off of regulatory demands on management attention.

In the 1980s, then, productivity will have some things going for it—not only a strengthening of positives, notably investment, but also a weakening of negatives that have helped retard productivity in the late 1960s and the 1970s. With a resolute effort to slow inflation, we can certainly reverse that scary-looking downtrend in the rate of productivity growth, and with an effective energy program that can help restrain the rise in world oil prices, we would have good prospects of seeing productivity rise at the projected rate of 2.1 percent.*

*Productivity did in fact begin to rise at a brisk pace early in 1981; but then came the recession, which inevitably caused a drop in the rate of productivity growth partly because the amount of idle plant capacity increased. For 1981 as a whole, productivity per worker in all private enterprise rose 1 percent. The quarterly rates of change were: first quarter, 4.7 percent; second quarter, 3.5 percent, third quarter −1.1 percent; fourth quarter, −7.2 percent.

The bad news is that even 2.1 percent wouldn't really be enough. With the projected 3.4 percent rate of growth in real GNP, output in the 1980s will fall short of meeting the claims on it. If productivity does no better than 2.1 percent, moreover, we will probably be jacking up our productivity less rapidly than our principal competitors in international markets. It will be hard to keep our competitive position from worsening, let alone to improve it.

So it would be well for the U.S. to strive to push its rate of productivity growth well above 2.1 percent in the 1980s. How far? A goal of 2.5 percent would not inspire anybody. Why not try to go all the way back to 3.2 percent?

That would be an exceedingly ambitious goal, to be sure. Two major factors that were pushing up the productivity growth rate in 1947–66—the shift out of farming and rising levels of education in the work force—will not have any lifting force in the 1980s. The 1947–66 period, moreover, enjoyed cheap energy, while in the 1980s productivity will be burdened with high energy prices.

Under these circumstances, a goal of 3.2 percent productivity growth might seem utterly unrealistic. But perhaps it wouldn't be entirely out of reach if the government and people of the U.S. seriously adopted it as a major national objective—an economic equivalent of war, as it were. The benefits of faster productivity growth would spread so wide that it might be possible to get a broad national consensus behind such a goal.

There are some obvious things a serious national program to improve U.S. productivity would include: tax changes to promote capital investment and research and development, as well as measures to make regulation more rational. But that would not be enough. To have any prospect of coming close to 3.2 percent in the 1980s, the U.S. would have to supplement conventional approaches with a special high-leverage program directed specifically toward productivity enhancement.

PROMOTING THE BEST PRACTICES

Dale Jorgenson, professor of economics at Harvard, indicates the direction in which we need to go. "What we have to do," he said

THE GLOBAL RETREAT OF GROWTH

The economies of the industrial nations have been hurt by high oil prices, but that doesn't entirely explain these deep dropoffs in productivity growth. Efforts to fight inflation have held down economic growth. And some fast-growth opportunities have been used up. But one thing hasn't changed: we were last in the 1950–73 period, and we still are. (Figures for the U.S. are lower than those in the text because a different measure of productivity is used in international comparisons.)

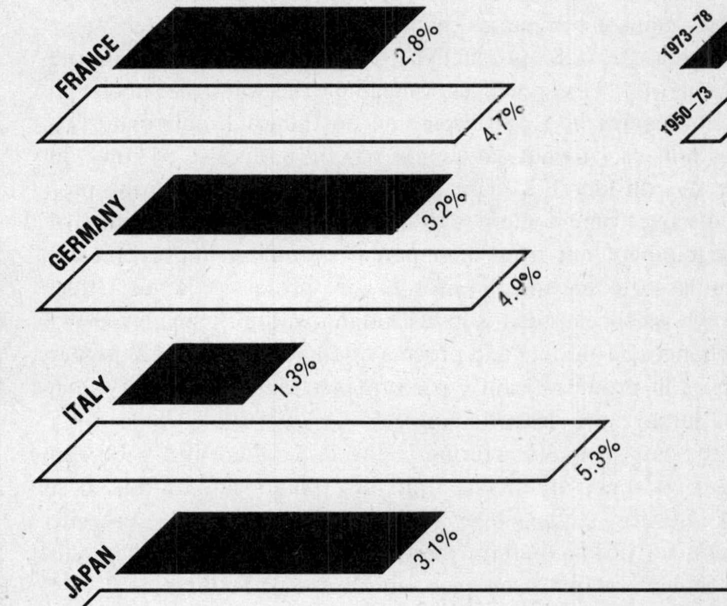

Annual rate of productivity growth

not long ago, "is develop institutions that bring technology in the U.S. up to the best levels quickly." The U.S. does not now have such institutions, and in this it is almost alone among the major industrial nations. Most of them have programs of some sort to promote the development and diffusion of advanced technology and the spread of good industrial practices.

What would U.S. productivity-enhancing institutions do? For one thing, they would encourage American companies using aged machinery or outdated technology to upgrade their equipment in large leaps. Economist Edward Miller of the American Productivity Center observes that the quickest way for an industrial company to improve its productivity is to "find out what the state of the art is and buy the best equipment." There is a lot of overaged or outdated industrial equipment in use in the U.S. today, and this lag creates an opportunity for getting extra productivity gains if companies catch up quickly.

An adequate U.S. productivity program would also promote the spread of "best practices," meaning the ways the most efficient companies in a particular line do things. Productivity is a matter not just of what equipment you have but also of how you use it. A considerable share of the gains in manufacturing productivity over time come not from development or acquisition of new equipment but from incremental shop-floor improvements, accumulating over time. Writes M.I.T. professor James Utterback, whose special field is industrial innovation: "Small step-by-step changes in product and process often add up to much greater advances in product quality, performance, and productivity than do the initial more drastic changes."

In any line of manufacturing in the U.S., there is a wide gap between productivity in the middling plants and in the best plants. Several studies over the years indicate that it is quite common for productivity in the best plants to be more than half again as high as in the average, and ratios of better than 2 to 1 occur surprisingly often. Sometimes the most productive plants do have the latest machinery or have advantages of scale or location that other companies cannot readily copy. But to the extent that part of the gap between best and average derives from transferable good practices, there are opportunities for catching up here too.

A national productivity program should also encourage company-wide productivity-improvement programs. A number of American companies that have adopted such programs, fitted them to their own circumstances, and stayed with them for extended periods have reported worthwhile results in terms of reduced costs and improved productivity. Their success suggests that many other companies, if they make serious and sustained efforts, can also achieve extra productivity gains over and above what they would get from investments in new plants, equipment, or processes. This is obviously a job for business, not government, but government institutions could help by providing reliable information and by sponsoring research—there is a need for rigorous, high-quality, objective research on what works, or doesn't work, under what circumstances.

KNOWLEDGE WITH LEVERAGE

The common element in the three aspects of this national productivity program is diffusion of knowledge, including the technical knowledge embodied in equipment. This diffusion of knowledge is where the leverage comes from.

Although much of the economy's productivity growth derives from growth in the capital-labor ratio, capital investment as such is not a very dynamic generator of productivity gains. For example, when a manufacturing plant adds capacity by acquiring more of the same kinds of machines it's already using, there is no obvious gain in productivity except for whatever improvement in efficiency is associated with the newness of the additional machines, and perhaps some economies of scale. When there is a large productivity-enhancing content in capital investment, it usually involves the introduction or diffusion of technological improvements of some kind.

This is one of the considerations that lead some of the most respected authorities on productivity, including Denison and Kendrick, to rank advances in knowledge above increases in invested capital as a source of productivity improvement. Both men use "knowledge" in this context as a broad category covering a variety of things—not just technological innovations embodied in new capital equipment but also managerial know-how and improvements in industrial practices.

Speeding up the diffusion of advanced knowledge has the same effect (for a while) as speeding up the advance of knowledge. It's true that rapidly filling in gaps and lags in knowledge will not impart any *lasting* lift to the long-term rate of productivity growth. Over a span of years, as the gaps that are worth closing are closed, we will exhaust the special opportunities to catch up. But in the meantime, we will have gained a lot: higher standards of living, higher levels of industrial efficiency, more resources for public as well as private use, a stronger competitive position in the world, and better prospects for the 1990s. It is hard to think of any good reason not to try.

December 3, 1979
Research associate: Peter Dworkin

2

THE NEXT INDUSTRIAL REVOLUTION

GENE BYLINSKY

An airplane is a plumber's nightmare. A DC-10, for instance, has three miles of hydraulic tubing that twists, bends, and turns throughout the airframe in triply redundant systems. Fabricating all that tubing used to involve a great deal of slow and expensive handwork. To begin with, it was necessary to construct a mock-up of an airplane's tubing, with craftsmen bending and fitting by hand as they went along. The mock-up sections would then be taken out and stored on warehouse shelves to serve as templates for subsequent pieces of tubing, which were also bent by hand.

Now master tubes are no longer stored on warehouse shelves. Instead, a designer routes replacement tubes from the computer memory where descriptions of them are stored. At the press of a button, the designer retrieves the key structural elements and then designs the needed section of a tube on the terminal screen with a fiber-optics light pen. At the press of another button, the needed section of tubing emerges from a computer-controlled tube-bending machine nearby.

The payoff from this new system has been remarkable. It's a lot more than just saving on warehouse space, although that in itself is substantial. The biggest benefit is that the tubes designed and fabricated under computer control fit into the airplane better, with fewer adjustments. When tubes were bent by hand for

F-15 fighters at McDonnell Douglas three years ago, as many as one hundred tubes per aircraft didn't fit. Now, using computerized tube-bending for F-18 fighters, McDonnell finds that only four tubes per aircraft have to be readjusted. As many as a dozen craftsmen used to bend tubes by hand for a single plane. Now the job takes three people: someone to watch over the tube-bending machine and two tube assemblers. Most valuable of all is the saving in time. Typically, under the old system it took Northrop Corporation six weeks from release of engineering drawing to bent tube in hand. Elapsed time now: eighteen minutes.

This is the world of CAD/CAM (Computer Aided Design/Computer Aided Manufacturing), where the future of manufacturing is being forged. If wisely and widely applied, CAD/CAM could overcome the productivity stagnation that has led to questions about the ability of American industry to remain competitive on the world scene. This modestly named technology, pioneered and developed in the U.S., has a power of explosive significance. Says a leading CAD/CAM specialist: "Revolution is an overused word, but that's what's going to happen here."

SAYING IT WITH PICTURES

The CAD of CAD/CAM is basically designing, drafting, and analyzing with computer graphics displayed on a screen. "Everything a draftsman needs is in the computer," says Ronald A. Cenowa, a computer engineer at General Motors' Fisher Body division in Warren, Michigan. "Anything that a draftsman conventionally does using triangles, pencils, compasses, and so on will be done mathematically within this system." CAD not only speeds up the slow and laborious work of drafting but also enables the designer to study various aspects of an object or assemblage by rotating it on the computer screen, separating it into segments, or enlarging or shrinking details.

What makes CAD so effective a way to design and analyze products and components is that the computer communicates with the designer in pictures. The mind absorbs the information content of a displayed diagram or drawing much faster than it can take in an array of numbers or words and translate them mentally into images. "Computer graphics," a scientist says, "seems to tap the way the brain is designed to work."

What's more, if the necessary programming has been done, the designer can analyze and test the things he designs right in front of his eyes, subjecting them to electronically simulated temperature changes, mechanical stresses, and other conditions that might impinge in real life. This on-screen testing can save the huge amount of time and expense involved in fabricating prototypes, then testing, modifying, and retesting. Donald D. Parker, a Fisher Body engineering executive, says in praise of CAD: "We do the design work on graphics and look at the results on graphics—all totally devoid of real parts to test. We can go through the whole process and have a good idea what the part is going to weigh, how strong it will be, how stiff, how well it will perform. We fix parts before the car is built."

Behind the visible magic of CAD lies a hidden base of programming. The power to design without setting pencil to paper and to analyze and test objects before they exist in the tangible world is created by vast, painstakingly put-together computer programs, often consisting of millions of lines of coded instructions. Battalions of programmers have labored to give the draftsman and the designer, at the press of a keyboard button or the flick of a light pen, the thousands of basic lines and splines (curves), and the immense fund of underlying calculations.

EASIER CHANGING IN MIDSTREAM

CAM, the other half of CAD/CAM, refers to something that has come to be commonplace in manufacturing plants—computer control of production machines. CAM-controlled systems range from machine tools running on punched-tape instructions to robots that can be reprogrammed to perform any of a variety of industrial tasks. CAM pays off apart from any connection with CAD, providing speed, accuracy, tirelessness, and dependability that human operators cannot match. But the potentials are greatly magnified when CAD and CAM are joined together.

When the linkage works smoothly, the on-screen designing and testing of a product generate a bank of computer instructions for manufacturing—or making the tools, dies, and molds used in manufacturing it. Even the tool paths, visible on the screen, can be specified. This CAD/CAM linkage greatly shortens the time between design and production. It is less costly to move to new models, to make midstream design changes, customize products, and set up short production runs.

Pratt & Whitney Aircraft, one of the companies that has gone furthest in applying CAD/CAM to manufacturing, now makes turbine blades, among other things, directly from CAD drawings, with the whole process automated. Edwin N. Nilson, Pratt & Whitney's manager of technical and management data systems and scientific analysis, says that in many cases, "Thanks to CAD we have gained a five-to-one or six-to-one reduction in labor and at least a two-to-one reduction in lead time. And these ratios go up as high as thirty-to-one and fifty-to-one where we linked CAD with CAM."

Someday, enthusiasts predict, CAD/CAM will make manufacturing into a process as smooth and as easily supervised as the flow of liquids in computer-controlled oil and chemical refineries is today. The march toward that goal began in the mid-1950s, when computers started to appear on factory floors in force to supervise the then-new numerically controlled (NC) machines. Developed under the sponsorship of the U.S. Air Force, such machines have since spread around the world.

In the beginning, "numerical control" referred mainly to the use of punched-paper tapes to guide metal-milling machines, but since then the concept has expanded to include many other varieties of machine tools, some of which are now directly controlled by computers without the use of tape. In flexibility, NC machines stand between fixed automation, exemplified by the presses that turn out automobile engine blocks, and the more adaptable industrial robots, which can replace human workers in forging, casting, painting, and even simple assembly operations. NC machines became an island of automation, albeit an important one.

TOWARD THE ULTIMATE VISION

Then, in the early 1960s, graphics terminals began to emerge—the dawning of CAD. Ivan E. Sutherland's Ph.D. thesis at M.I.T. in 1962 spurred the introduction of these terminals. Sutherland offered pioneering insights into the possibilities of computer graphics, and his paper convinced many people that CAD was a viable technology. Sutherland later went on to Salt Lake City, where he founded, with David C. Evans, Evans & Sutherland Computer Corporation, which makes highly sophisticated graphics terminals. With the spread of minicomputers and the continued decline in the cost of information processing, CAD caught on. Manufacturing engineers began to link computer graphics with NC machines. The production of hydraulic tubes in the aircraft industry is just one example of that link. NC machines driven directly by CAD also shape tools, fashion parts, and cut materials with laser beams.

CAD linked to NC machines is what CAD/CAM mainly means today, in its popular definition. That's an important link, but CAD/CAM theoreticians want to extend the concepts much further, moving toward the ultimate CAD/CAM vision, the automated factory. That involves not just connecting CAD terminals to computer-controlled machines, but also thoroughly computerizing a plant's or company's manufacturing operations, including control of the flow of parts and materials and the movement of products through the various stages of manufacture. There's plenty of room for improvement in factory efficiency. Studies indicate that in standard manufacturing a part spends about 5 percent of the time being machined and 95 percent moving or waiting.

A RESERVOIR OF WISDOM

Sometimes the label CAD/CAM implies full computerization, but the term has no precise and settled content—it means different things to different users. Accordingly, some experts in industrial computerization use the acronym CIM, for Computer Integrated Manufacturing, to refer to the concept of the automated factory. As the prophets of CIM envision it, movement toward full computerization of manufacturing creates a common reservoir of computer-stored wisdom regarding all the manufacturing processes of a factory, or a division, or even of a big multidivision company.

In an ideal CAD/CAM setting, many people in the factory should be able to tap the common data pool. Production planners and schedulers, accountants, and shop-floor foremen, among others, should all be able to plug into that store of manufacturing wisdom easily. The creation of such a common data base is the backbone of any system of computer-integrated manufacturing.

The achievement of something like full computerization of manufacturing entails a thoroughgoing analysis of all operations—and perhaps a restructuring of them—in order to translate them into precise computer language. This is a demanding and time-consuming task, and few companies are very far along with it. Even the most advanced practitioners of CAD/CAM—certain makers of electronic components, aircraft, autos, farm machinery, and machine tools—employ only a few of the building blocks of a complete CIM system, and for the most part as disconnected blocks. Factory operations may seem orderly enough until you try to describe them in computer programs; then they begin to look quite irregular.

Full computerization would embrace control methods that have been formulated in recent years. One is the curiously named "group technology," a means of coding and grouping parts on the basis of similarities in function or structure or in the ways they are produced. Application of group technology can enable a company to reduce the number of parts in use and to make the production of parts and their movement in the plant more efficient. Another key discipline is "process planning," which details the sequence of steps needed in the making of a part or product.

While some forward-looking companies make use of such disciplines, many others still don't. CIM seeks to computerize these instruction sets, as well as other manufacturing controls such as the planning of requirements for materials, tying them all together into a kind of computer superbrain that will run the factory.

Two organizations have been primarily responsible for focusing the attention of manufacturers on CAD/CAM: CAM-I (Computer Aided Manufacturing International, Inc.), a private group, and ICAM (Integrated Computer Aided Manufacturing), funded by the Air Force. Both emphasize integration. "We want to demonstrate," says ICAM manager Nathan G. Tupper, "that the power of integrating all the pieces is so much greater than the total of all the little bits and pieces. When you consider that sixty to seventy percent of the cost of manufacturing has got nothing to do with the physical making of a part, but has to do, in fact, with the planning, scheduling, and control of the equipment that makes the part and of the people who operate it, then it seems clear to me that that's where your emphasis should be. If you can build up a common information base and operate off that with intelligently constructed planning, scheduling, and control systems to drive your automated machinery, or provide instructions to your manually operated machinery, the payoffs are hard to imagine."

TOUGHER THAN GETTING TO THE MOON

There are obstacles to CAD/CAM, however, and some will be difficult to chip away. A major one is the lack of uniformity in procedures from one plant to another, or even from one technician to another in a given plant. Process planning, to cite one key activity, is still done largely with pencil and paper by individuals of varying experience; each one writes the plan to make a particular part somewhat differently. Another obstacle is the great diversity of incompatible computer software and hardware now in use in most corporations. One of the principal theoreticians of CAD/CAM, Dennis E. Wisnosky, who now directs the CAD/CAM effort at International Harvester, sees the integration task ahead as being maybe even tougher than sending men to the moon.

Among other things, the task will involve undoing the past. "In a valiant effort to get the job done," says Wisnosky, "manufacturing engineers continuously added complexity and controls. Then they divided the problem into small pieces, over and over again. As a result, many factories in many American industries are so complex that they appear to be unmanageable, labor forces seem to be out of control, and costs are all but unknown." In time, he adds, the development of CIM will cure those ills.

Wisnosky is putting his ideas into practice at International Harvester, where he has assembled a large group of specialists. Currently beset by financial troubles, the company sees CIM as a lever that will help it vault back to the top of its field. Wisnosky and his colleagues work with rotating groups of manufacturing engineers. They began with the idea of devising an architecture, or highly detailed plan, for CAD/CAM. The purpose is to describe fully, in programmable terms, the functions of manufacturing and the dynamic interaction of all the subsystems.

"The recognition that we can attach everything to everything else is somewhat overwhelming," says Margaret A. Eastwood, manager of CAD/CAM architecture and simulation at International Harvester. "And I think that's why you see it being implemented only in little clumps—because it's very difficult to keep it all straight. But it's exciting." Wisnosky and his teammates model and test planned modifications in International Harvester factories prior to any major investments in facilities or changes in computer software. The aim is to eliminate unpleasant surprises in putting CIM into place.

This is CAD/CAM modeling on a grand scale. It not only examines materials, tooling, equipment, and parts geometry for each process, but also seeks the most economical way to do the job. "Our objective," says Wisnosky, "is to match the best process with the best material at the best time."

THE HOT-SHEET EVIDENCE

The integration, in effect, seeks to devise a computer-aided scheduling system in which computer controls would respond and react to changes on the factory floor to keep that schedule running smoothly. "There is no system today that can handle this on the factory floor," says Wisnosky. "In every plant in the country you see foremen running with hot sheets in their hands—lists of parts that are needed immediately on the assembly line. Some plants have more expediters than people making things. That's where the tremendous impact of CAM will be in the future—that whole area has not been tapped."

Fortunately, a company need not wait for the full development of CAD/CAM potentials. As a growing number of corporations are discovering, the available elements of CAD/CAM can already start paying impressive dividends in increased production, in time and money saved, in better-quality products, and in greater flexibility in manufacturing. Using some CAD/CAM systems, GM, for example, turns out car-body components such as fenders in half the time it took with manual techniques. What's more, the shapes and dimensions of components produced the new way are more accurate.

Even CAD by itself is a phenomenally powerful technology. At Fisher Body, for instance, a car made of electronic mesh looks like a blueprint in motion as it bumps along a computer-simulated rough road on a terminal screen. The car's body undulates in a purposely exaggerated fashion when it hits a make-believe bump so that engineers can spot design flaws before prototypes are built. In a similar type of test at International Harvester's engineering-design center in Hinsdale, Illinois, an electronic image of a combine, showing how it would contend with bad weather conditions on the farm, spits out columns of indicator numbers as it strains to extract its wheels from simulated mud on one side and slithers over simulated ice on the other.

If a designer is not fully satisfied with the initial creations, he or she can try many variants, taking chances with computer data that could never be taken with an actual product. Thanks to this ability to improve products quickly even as they are being designed, says a CAD/CAM executive at General Electric, "a whole new era of producibility" is beginning.

Being able to try out an almost infinite variety of materials, product configurations, and other factors in manufacturing products before committing actual resources can bring large gains in efficiency and substantial reductions in costs. Roland W. Schmitt, vice-president for research at GE, reports that savings just from eliminating mistakes in the making of metal molds for plastic parts (GE produces more than 130 million pounds of plastic parts per year) will soon amount to $100 million a year. Images on a terminal screen can simulate the cooling of a plastic part inside a mold, for example. If the simulated part cools unevenly, a designer can readily see the unevenness, indicated by color variations.

Even more capable CAD is on the way as the new computer-graphics technology emerges. Miraculous as today's CAD may seem, it has certain shortcomings. The images usually consist of a network of lines that looks something like wire mesh. This type of structural representation describes the outlines of the object mathematically. Such basic properties of solid objects as volume and center of gravity, for instance, cannot easily be calculated. In wire-mesh images, moreover, interaction between complex parts is sometimes difficult to discern.

THE CLAY IN THE COMPUTER

An ideal CAD technology should produce more realistic representations of solid objects. The new three-dimensional modeling does that, and it will drive CIM forward on a broad front. Besides making the design and analysis of complex parts much easier, it will provide vital information to everyone in the factory with greater clarity and precision than is possible with wire-mesh modeling. A process planner, for instance, should have a clear, three-dimensional conception in his mind, not only of the completed part but also of the changing shape of a block of metal being worked on as it proceeds through the machining process. Solid modeling can give him that as a motion-picture-like sequence of images.

With solid modeling, computer programs can represent internal structures as well as external shapes. A designer can construct representations of complex parts the same way a craftsman would make a model of clay, except that now the clay is information stored in the computer. When the electronic solids are cut apart on the screen, all the details one would find inside a real part are clearly visible—a boon to product designers and engineers.

What this means is that the design and analysis of complex parts and tools will become much easier than with wire-frame graphics. Later on, the parts will be manufactured directly from designs. That will represent a considerable expansion of CAD/CAM capabilities, which heretofore have been pretty much limited to objects with fairly simple internal structures.

Little wonder, then, that knowledgeable people view solid modeling as the most important single development for the future of CAD/CAM. Norway, which hopes to use CAD/CAM to free workers from the need to toil on third shifts, has traded millions of dollars' worth of natural gas as part of the price of participating with West Germany in a $30-million CAD/CAM program in which solid modeling plays a central part.

Solid-modeling systems are just beginning to be commercially available. A system from Applicon Inc. of Burlington, Massachusetts—really a complex computer program—sells for $50,000. Other solid-modeling programs are being developed; perhaps the most sophisticated one is being worked out at GM's research labs for in-house use.

SEEING INSIDE THE ENGINE

Applying programs developed in collaboration with the University of Rochester, GM technicians found they could use solid modeling to design a car trunk so as to get the most storage capacity. This seemingly simple chore had in fact been pesky and time-consuming; it required the building of full-size wooden mock-ups. GM researchers have also applied their solid-modeling system experimentally to such complex problems as measuring the moving flame inside a simulated auto-engine combustion chamber—very much akin to looking inside a working engine. What's more, the system makes it easy to vary the chamber design and study the effects of such changes on the combustion process.

A multitude of commercial packages combining hardware with software and offering some unified CAD/CAM capabilities are available today from such vendors as Applicon, Computervision, Calma, and others. (GE acquired Calma in 1980; Schlumberger is on the way to acquiring Applicon.) Prices range from about $150,000 to $800,000 and more. The CAD/CAM vendors may constitute the fastest-growing industry around, an industry expected to hit $750 million in sales this year and $2.2 billion by 1984. It is Wall Street's latest darling, and its price-earning ratios reflect that infatuation.

Using CAD/CAM packages, which usually pay for themselves within a year or so, can bring pleasant rewards. One happy customer is Robert M. Ronningen, president of Ronningen Research & Development Company of Vicksburg, Michigan. His company makes a wide variety of plastic molds, parts, and prototypes for manufacturers of various products, from steering wheels to computer terminals. A little over three years ago he installed a Unigraphics minicomputer-based CAD/CAM system, marketed by McDonnell Douglas Automation Company, a subsidiary of the aircraft maker. Since then Ronningen Research has more than tripled in size, from thirty employees to one hundred.

"I STILL GET KIND OF AMAZED . . ."

A mechanical engineer who says that he is "always looking for a better way to build a mousetrap," Bob Ronningen credits his CAD/CAM system with lifting his company to the forefront of prototype manufacturers. "I still get kind of amazed how fast we can do things now," he says. "We can build an injection mold in two to three weeks with Unigraphics. Without it, the same work would take from eight to sixteen weeks."

At Ronningen Research CAD/CAM takes a product from design to manufacture under computer control. The merger between CAD and CAM is much more difficult to execute on a large-factory scale, but more and more corporations are trying. Some, notably in aerospace, have already carried the integration a long way. McDonnell Douglas, one of the pioneers, has devised a common data pool that is directly accessible to aircraft-structure analysts, design engineers, numerical-control programmers, and parts inspectors. The output, in the form of CAD drawings or numerical instructions, speeds along special communication links to automatic drafting machines, metal-milling tools, and inspection devices. Control programs monitor and direct all these busy pieces of apparatus. A management data-acquisition system watches over the status of all machine tools, and frequent status reports inform shop managers about the number of completed units, unit costs, and operator productivity. Today McDonnell makes about 85 percent of a fighter plane by automatic means. Yet an executive concedes that even McDonnell has not yet reached the CAD/CAM "destination."

Reaching that destination will entail the further step of computerizing factory-control functions. Lockheed Georgia Company is getting there. It took a program for computerized process planning, developed by CAM-I, and improved on it to a point where a process planner now can write fifty to one hundred simple production plans a day, compared with the three or four he used to do writing by hand. But to get that logic into the computer, Lockheed Georgia grilled the best of its manufacturing people for seven months. This is another piece of evidence that to reap the full benefits of computerization, manufacturers have to look very deeply and painstakingly at their operations in order to understand them well enough to make them understandable to computers.

Smaller companies will have to rely on programs they can obtain from vendors, and there's plenty more help on the way. Some big guns are moving into the CAD/CAM marketplace. GE has made a $500-million commitment to become a "world supermarket of industrial automation" in the 1980s. Senior vice-president Donald K. Grierson has promised that GE's systems will be able "to blast productivity through the roof."

TO BE WITH IT AT THE CLUB

IBM, which for some years has been offering a Lockheed-developed CAD/CAM system that runs on IBM computers, is highly interested in factory automation as an expanding market. Both Sperry Univac and Honeywell also have plans to offer big CAD/CAM systems that will go beyond the smaller versions offered by Computervision, Applicon, and other independent vendors. The mainframe makers feel that advanced solid modeling, in particular, will demand huge number-crunching capability.

With big computers or small, CAD/CAM is moving along. Joseph F. Engelberger, president of Unimation, Inc., the leading maker of industrial robots, declares, "The word is out" that CAD/CAM is the wave of the future in manufacturing. "I don't think a guy will be able to go to his country club if he doesn't have a CAD/CAM system in his factory. He's got to be able to talk about his CAD/CAM system as he tees off on the third tee—or he will be embarrassed." He may be worse than embarrassed if CAD/CAM prophets are right that it offers the only means by which the U.S. can survive as a competitive industrial power.

The mystery factor, as usual, has to do with the Japanese. Will they grab a technology that is as American as the space shuttle and run away with it, as they have done with so many other technologies? Right now the Japanese are lagging in linking CAD with CAM, but they are catching up fast, mainly by buying American technology. They can hardly have failed to hear the refrain that CAD/CAM vendors and technicians never seem to tire of repeating. "CAD/CAM," it runs, "has more potential to increase productivity than any other development since electricity."

October 5, 1981
Research associate: Alicia Hills Moore

3

SUCCESS AT THREE COMPANIES

EDWARD MEADOWS

Problem: How to keep your Whoppers really hot without warming the iced drinks stationed next to them. When a Burger King restaurant manager in Des Moines came up with the answer by sculpting a $25 piece of sheet metal to concentrate heat in the Whopper bin, he made a major contribution to the Whopper Hotness Program, part of the productivity drive at Burger King's Miami headquarters. Food researchers there had painstakingly calculated that, *ceteris paribus,* hotness is what sells Burger King's hamburgers, and they set about calibrating that hotness to a precise 130 degrees. A consistently hot Whopper, plus ice in the drinks, means more sales per hour, which in the fast-food industry is how you measure higher productivity.

Burger King can't do much all by itself to reverse the alarming trend in American productivity. But if it's true that God is in the details, more often than not so is productivity, and corporate managers are looking at the tiniest details these days. Many of them see productivity improvement as the only way to keep ahead of the pack in an era when they can't easily pass on cost increases without losing some of their market share. Collectively, the efforts of all these companies may add up to quite a lot.

The kinds of changes that pay off vary widely from company to company, of course. With this in mind, *Fortune* sampled unusually successful productivity programs at three very different corporations. Burger King is a fairly typical service company; Corning Glass Works is a high-technology, energy-intensive manufacturer; and Crompton Company, the second largest maker of corduroy in the U.S., is a successful competitor in a low-technology industry.

Businessmen see productivity through a wider lens than is customarily used by economic analysts. On the factory floor, almost any sort of improvement may be counted as a productivity booster. It isn't one, however, unless it increases output by more than the increase in input—or reduces the amount of input needed to produce the same amount of output. When output and input are counted in dollar terms, this latter technique becomes manifest as a cost-reduction program. In fact, it is by way of cost-reduction programs that most corporations tackle productivity.

CORNING'S COST CURVE

One of the most sophisticated such programs in American industry is in force at Corning Glass. Throughout Corning's appropriately glassy corporate headquarters building in upstate New York, the word productivity is on the tip of almost everyone's tongue. In their zeal to spread the doctrine, the company's executives are apt to draw Corning's learning curve on the back of an envelope. That curve is a downward-sloping line illustrating how the cost of making glass has fallen as the cumulative total of units produced has risen over the years. In 1880, when Thomas Alva Edison asked Corning to make him up a batch of glass envelopes that he could fashion into light bulbs, glassmaking was a handicraft. By 1926 a "ribbon" machine was invented to replace hand blowing, and Corning's cost curve dropped dramatically. More recently Corning developed an all-electric furnace, which has been responsible for yet another large downward swoop in the curve.

Important as those big improvements are, nonetheless it is small innovations that keep the curve coursing in the direction of lower costs. The company's planners make a guess at what inflation will be for the next five years and then attempt to keep their own costs under control. "I'll be darned if I can figure out how we're going to do it," says Thomas C. MacAvoy, president of Corning. "I think, as in all technically innovative businesses, frequently you don't know what new inventions will come along, so you don't know how in heck you're going to achieve the goal." So far, Corning has done well. Productivity in the glass industry has been growing at a 3 percent rate throughout the 1970s, but Corning has got its productivity growth up to 6 percent—the goal it set for itself after an eye-opening company study in 1970 revealed that it was doing no better than the industry as a whole. MacAvoy says most of the gains have come from technology. "We are always trying to redesign jobs, motivate people, and whatnot, but frankly I don't think people are basically lazy and ineffective. I don't think it's possible to get them to work twenty percent harder. To get them to work smarter is different, and that involves the application of technology."

AN END TO BLEMISHES

A tour through the Corning Pressware plant, which makes laminated-glass dinnerware, reveals the sort of small improvements the managers have devised. Next to one of the long kilns in which the dinnerware is fired stands a plain white cubicle that will soon become a centralized control panel. Because the kiln operators will be working together rather than at separate stations, they will be able to control the flow of production more smoothly, nudging productivity up a few points.

Farther down the line, the molds for the dinnerware yielded another kind of improvement in efficiency. Their vacuum holes, which adhere the melted glass to the shape of the mold, were changed from pinholes to slots. This change strengthened the plates, got rid of the tiny blemishes created by the pinholes on the bottom of the plates, and made possible a speedup in the pace of the production line.

Right in the center of the plant are three big sets of traffic lights hanging from the ceiling. A green light means a production line is running smoothly; amber advertises the fact that a line needs close inspection because defects were found in some of its dinnerware. Red means a shutdown. The signals are regulated by quality inspectors, who used to check each plate but who now take a sampling. The new method is much faster but also has led to better quality control.

The Pressware plant, like all thirty-nine Corning plants in the U.S., has its own productivity coordinator. Usually an engineer himself, he works closely with the plant's engineering staff and can call in help if needed from the corporate manufacturing and engineering divisions. But attention to productivity doesn't end with the engineers. Ronald Matthews, the plant manager, has to come up with a zero-base budget each year, and he has to analyze where "cost opportunities" lie for the next five years.

Cost opportunities for every plant go into each division's cost-reduction "portfolio." In fact, only when a plant has firm cost objectives and cost-reduction plans is it said to be "under management." Coming under management is a process that brings the manufacturing people together with marketing men and financial executives to iron out cost reduction and pricing. This is a formal program, and it gets complex, with charts showing how every expenditure and price increase affects the gross margin.

Sometimes getting a unit under management can spell its demise. John Rudolph, planning manager for Corning's manufacturing and engineering division, tells of the company's experience with Dur-Cor, an exotic coating used in the manufacture of electronic components. "Since it wasn't doing well at all in the marketplace," says Rudolph, "the managers sat down and said something's got to give here. They looked at what had to be done in terms of cost reduction and at the capital spending that was required to do it. They decided it ain't worth it." Corning got out of the Dur-Cor business.

White-collar productivity is also crucial at Corning, since the corporation depends heavily on bright, innovative scientists and engineers. The company hopes to improve the productivity of the corporate engineering people by bringing them together under one roof in a new building. The coffee lounges will be equipped with blackboards to encourage the engineers to talk out technical problems informally among themselves. Escalators will be installed rather than elevators, to facilitate movement from floor to floor and thus make face-to-face communication easier.

Corning's productivity system is pretty advanced, but Rudolph thinks it is crude compared with the new system he hopes to see installed by 1981. Once the company updates its computerized accounting system, Rudolph and his staff want to incorporate programs to measure regularly the total factor productivity at each plant. This will tell managers exactly how much productivity they are squeezing out of labor, capital, energy, and materials. "Labor productivity alone masks too many things," explains Rudolph. "It hides the trade-offs." In some cases, he points out, mechanizing a production line can cost more than it's worth.

The more refined productivity data should be especially helpful to Thomas Howitt, Jr., Corning's energy expert. "Our total energy cost in 1972 was $18 million," he says, "but between then and 1978 our volume grew and so did the price of energy, so that if we had been operating the same as in 1972 our energy bill would have been $80 million in 1978." Instead, the company ended up paying only $56 million that year.

The resulting $24 million of "cost avoidance" is no mean sum to Corning's executives. It was achieved mostly through conversion of glass-melting furnaces from natural gas or oil to electric firing, which is three times as efficient. To hasten the conversion, Corning developed its own patented electric furnace. More energy was saved by such measures as removing alternate fluorescent bulbs in offices and watching room temperatures closely. Still, Howitt feels his job has just begun. As energy costs continue to rise, new ways of saving energy will become worthwhile. "We will have more plums to pick off the tree," he says, "plums which were not at all attractive to us in the past."

PUSHING THE MACHINERY

Much in contrast to Corning's sophisticated technology is the relatively uncomplicated and old-fashioned machinery used by

Crompton, a New York–based manufacturer of corduroy and velveteen. Crompton has six plants in the South and is one textile company that isn't being beaten to death by cheap imports; instead Crompton aggressively exports its fabrics throughout the world. This exporting strategy is not a cause but an effect of Crompton's productivity. In order to push the efficiency of its weaving equipment as high as possible, the company keeps its machines running twenty-four hours a day, 350 days a year. New markets provide the customers to buy the extra output produced by the nonstop operation. Of Crompton's $158 million in sales in 1979, more than $52 million came from overseas.

"In my experience," says William Lord, Crompton's president, "ever since the time of the Korean war you couldn't survive in the textile business in this country unless you were getting pretty close to the most out of your equipment. It's so damn competitive." In this industry, a plant running at 88 percent of capacity is considered inefficient, while one going at 92 percent is doing fine. The 4 percent range doesn't leave much room for error.

Still, Bill Lord is leery of rushing to buy the newest generation of looms, which are highly productive yet hard to reprogram. He believes Crompton has to stay flexible enough to suit the changing tastes of his high-fashion customers. To compensate for less efficient machinery, Crompton's executives must work harder to extract greater productivity.

THREE-DAY WEEKS

The overriding ingredient in the textile business is people. Though many of the jobs are semiskilled, the way workers perform them determines the quality of the fabric. In addition to the low-skilled employees, there are some whose tasks approach an art form, such as the men and women who watch over the cutting of "ribs" in corduroy and velveteen. One wrong move and a worker can tear yards of fabric into wasted shreds that have to be sold for pillow stuffing. Even in the best-run plants, a lot of imperfect goods come off the production line. It would be hard to imagine an auto plant where more than 10 percent of the cars turn out to be defective, but it is not unusual for that percentage of the finished cloth to be rated second quality. It has to be sold at price discounts of up to 70 percent, if it can be sold at all. Holding rejects to under 4 percent, as Crompton does in its most efficient mill, is considered an accomplishment.

At the Crompton plant in Leesburg, Alabama, the latest productivity wrinkle is a three-day, twelve-hour work schedule that gives the mill hands a full week off once every eight weeks, as well as four days off every working week. Other textile companies also use multiple shifts to keep their plants running around the clock, but usually the arrangements are harsh on the workers, with constantly rotating schedules that result in heavy absenteeism. Crompton's plan has the extra attraction of bonus pay for employees who stay on the job; they are paid for forty hours if they complete the required thirty-six. While the schedule sounds grueling, Lord says his studies show that the last few hours of the three-day week are actually the most productive.

Crompton's work arrangement is a boon to the many employees who have small farms to tend. It has held the rate of labor turnover to 9 percent a year, an impressively low figure in an industry where the average rate is over 50 percent. The Leesburg plant has a backlog of applicants for any jobs that come open, and the workers voted in a bargaining election to remain nonunion. Impressed by the low training costs and high productivity at Leesburg, Crompton is installing the new work schedule in its Arkansas and Georgia plants.

A bit of computerization has added some competitive spirit at Leesburg. With the push of a few buttons, the loom operators can check up on what their output is and compare it with that of their cohorts. Plant officials say that the workers do check up on themselves, often cutting a lunch break short to stroll by the terminals for a readout.

Crompton does its share of engineering to improve the efficiency of its equipment, but has found that the results sometimes aren't worth the effort. For instance, it tried out a complicated new piece of equipment to replace the workers who change the bobbins on spinning machines, but Bill Lord noticed that the contraption had to be constantly attended by three engineers, so he got rid of it. The payoff was decidedly more satisfactory when Crompton's engineers found a replacement for the spring-steel blades used to cut the tiny grooves that give velveteen its plushness. The spring-steel blades didn't keep a sharp edge, so the engineers substituted ordinary Wilkinson razor blades and got a 60 percent boost in efficiency.

Company-wide, the emphasis on productivity at Crompton has yielded an annual increase of 8 percent in output per worker-hour since 1974—well above the average for the U.S. labor force as a whole. Given the high utilization of Crompton's capital equipment, the increases in efficiency would be even more impressive if they were figured in terms of total factor productivity.

JUSTIFYING BREAKFAST

Burger King's bright, sunny corporate headquarters in Miami bubbles with talk about "justifying breakfast on a capital-cost basis" and restaurants that have a "disco capability." Burger King has got hamburger-making down to a science, and the essence of that science is productivity. The conventional wisdom is that productivity is hard to achieve and hard to measure in service industries such as fast food, where quality is important. Burger King disagrees. Everything it does is measured and fed into computer simulations.

Burger King's executives like to say theirs is a company run by 50,000 teenagers, and those kids are getting more and more expensive. "We've had a lot of scares about beef prices," says Donald Smith, Burger King's president, "but the truth is that labor costs have risen at a much faster rate. We see the pressure to increase the minimum wage continuing because it seems very popular among the politicians."

To fight these higher labor costs, Burger King looks equally hard at its people and its machines. Each Burger King restaurant is thought of as a small food factory. There is, however, one salient difference between a fast-food factory and other assembly lines: an increase in production translates immediately into extra sales rather than into inventory. "The demand at the peak hours is so great that if we can produce more we can sell more," says Donna Nicol, Burger King's communications director. "We have to do it within a very tight time frame," she says. "Nobody cares how many hamburgers we can make between eleven p.m. and six a.m."

Every movement of each employee in a prototypical Burger King restaurant is calculated and readjusted in time-and-motion studies that are over a half-century old by now but seem to have lost none of their validity. For example, at drive-in windows, just moving the bell hose—which triggers a loud ring when cars drive over it—can make a big difference. As members of the company's Drive-through Task Force found out, order takers needed eleven seconds to react when the bell hose was located near the drive-in window. The productivity experts moved the hose back ten feet, so that by the time a car had braked, the order taker was waiting to scribble down what the customer wanted. With this and other small changes, the company has been able to cut the order transaction time to thirty seconds. That means the drive-in window can handle an extra thirty cars an hour.

To serve those extra cars, the chain has to have more food coming off the production lines inside, so the engineers at Burger King installed computerized french-fry machines. They are putting television terminals in the kitchen, so that the chefs can read incoming orders off the screen, and the company's engineers have visions of machines that will mix, pour, and cap soft drinks automatically.

All this is first tested on a computer model of a typical Burger King restaurant. If it passes there, the scientists try the new wrinkles at one or more of the five research-and-development restaurants in the Miami area. The r & d restaurants, which do $5 million of business a year themselves, provide real-life testing labs for the company. They serve unsuspecting customers with the latest refinements in hamburger technology.

Much of the impetus for productivity improvement comes from the specialists at work in such team efforts as the Drive-through Task Force and the Whopper Hotness Program. But plenty of good ideas come from individual Burger King store managers, who get cash rewards when they turn in useful productivity suggestions. On the other hand, when the company devises a neat productivity-enhancing plan that calls for a bit of capital spending, the onus is on the company to sell the idea to its restaurant managers, the majority of whom are franchisees. (Burger King likes to own and operate at least one-fifth of the units in an area so the corporation can get a feel for the market.)

DECISIONS IN A DAY

When Burger King puts up its own restaurants, land and building costs now run to $500,000 a unit. Ronald Petty, Burger King's real-estate chief, looks hard to find ways of cutting costs, and with some success; he says he saved the company roughly $1.4 million in 1979. By using computers stocked with predigested demographic information, he has cut down the average time spent on site approval from two weeks to a seemingly incredible one day. By putting construction specialists in the company's ten regional offices and making sure they know all about their areas' various building codes, he pared the time spent on design of each outlet from six weeks to two. Before this program got under way, Burger King had five different restaurant plans. Petty eliminated all but two—one for small towns and the other for larger-volume areas. Just teaching the regional staffers how to negotiate real-estate deals has saved plenty, too.

The immense amount of constant fine-tuning that goes on at Burger King has helped bring the corporation neck and neck with its archrival, McDonald's. A lot of the improvements have been made since 1977, when Don Smith moved over from McDonald's to take charge. Since then, Burger King's per-restaurant sales have grown apace with McDonald's. The operating margin is still 12 percent versus McDonald's 18 percent, but Burger King is catching up; it is already Pillsbury's fastest-growing division.

Efforts like these seem of tremendous import to the country's burgeoning band of productivity specialists, such as C. Jackson Grayson, Jr., chairman of the American Productivity Center in

Houston. They say that if the best technology and best practices of the most efficient companies could be widely and rapidly diffused, U.S. productivity would make an impressive spurt, closing much of the "productivity gap" that opened in the 1970s. They also point out that this new attention to productivity amounts to nothing more than getting back to basics.

There have always been two fundamental ways to widen business profit margins. One is by increasing demand for a product. The other is by reducing unit costs. During the ebullient 1960s, when horizons seemed boundless, corporations put much of their effort into marketing. Now, as the examples of Corning, Crompton, and Burger King attest, corporate managers are paying increasing attention to the supply side of the equation. Indeed, in an environment of stagflation, many corporate executives say productivity is really what the game is all about. Getting good at it is also, nowadays, one of the most promising ways to get to the top corporate office.

March 10, 1980

4

REDISCOVERING THE FACTORY

ROBERT LUBAR

Under the benevolent rule of the Reagan Administration, business will probably get most of what it has been asking for in tax and regulatory relief as well as more attractive incentives to invest. But what business does for itself—specifically, what it does to correct one particular management oversight—could turn out to be far more decisive than government favors in restoring U.S. industrial vigor and competitiveness.

A good many corporate managers, shocked by faltering productivity and loss of markets to Japanese and other rivals, have begun to perceive a connection between their setbacks and their neglect of the part of their business that actually produces the goods. They have been reminded that in production the *how* can be as important as the *what,* and they are going back to basics—or, in the metaphor of John A. Young, the CEO of Hewlett-Packard, "back to blocking and tackling." As Young observes, "There is a feeling in the business community that managing the fundamentals well is something that has escaped us a bit."

This rediscovery of the factory has been taken up as a theme and given systematic formulation by the redoubtable intellectual infrastructure that has grown up around business—the faculties of the leading graduate business schools and the management consultants. Issues involving production management (or "operations management," as it is often labeled so as to encompass the "production" of services) are popping up on the contents lists of academic journals, on the agendas of executive forums and seminars, and in after-dinner oratory. Indeed, one can sense a formidable missionary movement building up.

Its chief prophet is Professor Wickham Skinner, an astute and articulate analyst in the production and operations management department of the Harvard Business School. Skinner's ideas first won a wide audience in 1976, when his article "Manufacturing—Missing Link in Corporate Strategy" appeared in the *Harvard Business Review.* Management consultant firms such as Arthur D. Little and Booz Allen & Hamilton invited him to conduct seminars for their key people and then introduced aspects of the Skinner approach in counseling their clients. A few corporate executives made the pilgrimage to Cambridge to hear the message directly.

THE HEROES WERE CAST ASIDE

The gist of Skinner's case, peppered with a dash of polemics, is that "manufacturing is generally perceived in the wrong way at the plant level, managed in the wrong way at the corporate level, and taught in the wrong way at business schools." Things began to go awry in the late 1950s, around the time John Kenneth Galbraith was confidently announcing, "We have solved the problem of production." Up to then, many U.S. industrial companies were run by production men. Their genius at getting the goods out was credited, in large part, with winning World War II and powering the postwar surge in economic growth. At the peak of their success, however, the production heroes were shunted aside. With rapidly proliferating new products competing for the consumer's favor, marketing became the chief challenge for management and the favored route to executive command. A few years later, with the coming of the conglomerates, big mergers, and battles over tender offers, and the rising importance of money management, the lawyers and financial men began to dominate the scene.

Production became isolated from the corporate mainstream. It was taken for granted and left to the specialists: the technicians in the plant who could be counted on to fill orders. They in turn were not expected to concern themselves with such things as return on investment and corporate strategic planning. They tended to think in terms of tonnage produced rather than dollars earned.

Companies generally recruited their factory managers from engineering schools or trained people with simple college degrees. Production jobs were shunned by the MBAs who were streaming into professional management. They had no desire to sweat their lives away in grubby factories out in the sticks, far from where the action was. Impatient to run the fast track to the top, they had no time to get firsthand knowledge or experience in manufacturing.

As a consequence of the communications gap that developed, corporate headquarters was unsympathetic to—or unaware of—the factory manager's problems. When trouble showed up, top management's inclination often was to get rid of the business instead of looking for the causes.

Skinner discerned the full and unfavorable consequences this had for overall corporate performance. Top management was making strategic planning decisions without taking into sufficient account the capabilities and limitations of the manufacturing operation. In reality, production cannot be taken for granted. There is no single most efficient way to make everything. "A factory can only do certain things well," Skinner notes. "There's a constant trade-off, for example, between cost and quality, between short delivery times and controlled inventory. When you recognize that everything is a trade-off, you can tell the factory what's most important for it to do. The factory manager has to get dealt in on the company's strategy so he can make structural decisions that match the strategy. If the manufacturing is right, it's a competitive weapon; if it's wrong, it's a corporate millstone."

To illustrate this last point, Skinner cites an episode that took place at Honeywell, where he worked for ten years before joining the Harvard faculty. One Honeywell plant was devoted to making gyroscopes for highly specialized scientific and technical use, and fuel gauges for airplanes. The two production lines were intermingled on the factory floor, and eventually trouble developed. "Gyroscopes were ten times harder to make," Skinner recalls, "but the Honeywell people were having trouble competitively with the fuel gauges. They did everything to try to figure out why the costs couldn't be kept down. They made accounting analyses, hired an MBA to come on board. Nothing worked, so they decided to get out of the business.

"Then one of the managers whispered a suggestion to the plant boss, who asked top management for $20,000 and three months more time. With the $20,000 they bought plywood and some two-by-fours and walled off a corner of the factory. They put everything that had to do with fuel gauges in that corner and segregated the workers. Within six months the problem was licked."

The moral Skinner draws from this experience is that the manufacturing mission should be clearly focused. In the sense that he uses it, "mission" is descriptive rather than inspirational. It describes the whole production process, from the sources of raw materials to the delivery of finished goods, embracing such things as the extent of vertical integration, the degree of automation, the length of production runs, and the product mix. Obviously, the manufacture of gyroscopes and fuel gauges involved entirely different missions—different kinds of workmanship, a different production pace. Mixing the two turned out to be bad for the fuel gauges.

Skinner stresses the importance of integrating the manufacturing mission with the corporation's competitive strategy, so that they are mutually consistent and reinforce each other. This was illustrated by the case of one Booz Allen client: S.C. Johnson & Son, the company that makes Johnson's Wax. In the past it had been mainly a producer of brand-name chemicals for household use—insecticides, furniture cleaners, floor polish—but in recent years it has ventured into industrial products and personal-care items such as shampoo, and its future growth seemed to lie in those areas.

The company recognized that it had to adjust its manufacturing system to the new marketing direction. Up to then it had concentrated all of its U.S. production in a single highly automated, low-cost plant in Racine, Wisconsin. As new products were developed, more machinery and employees were crammed into the plant. That was satisfactory as long as Johnson's product line was fairly homogeneous. But as it grew more diverse—and operations more complicated—problems threatened. Moreover, the firm feared losing personal touch with its employees if the plant got too big.

After studying the company's entire business and its marketing strategy for the future, Booz Allen came up with a long-range manufacturing plan. Instead of concentrating everything in Racine, Johnson was advised to rely more on outside contractors to do its manufacturing and build some new plants around the country. Each factory would focus on what it could do best—Racine on high-volume, dependable sellers like Raid insect killer; the smaller satellite plants on items with short production runs that had to be constantly attuned to fickle demand in the marketplace.

The recommendations were promptly accepted by the company. "Essentially there wasn't anyone who disagreed," says Dick E. Milholland, Johnson's vice-president for U.S. manufacturing. "There's no doubt this is a market-oriented company, but this new strategic approach is a major manufacturing decision."

PROFITING FROM QUALITY

When a company effectively harnesses its factory planning and management to corporate objectives, it is likely to do better at raising productivity, developing and assimilating new technology, improving product reliability, and applying the new quality-of-worklife concepts in its labor relations. All these endeavors call for a total corporate commitment, from top management all the way down.

Until fairly recently, productivity was an economists' term, rarely used by businessmen. Now it's one of the most popular words in the business lexicon, though corporate executives are no nearer than economists to penetrating the intricate chemistry that makes productivity grow or stop growing. It is dawning on some managements, however, that poor productivity performance cannot be blamed simply on the decline of the work ethic, unreasonable government regulation, or even a shortfall in capital investment. They are beginning to see it as a symptom of something endemically wrong with the way their manufacturing operations are set up and organized. Good productivity has to be built into the factory from the ground up.

The same holds for product quality. The tendency has been to blame errors and defects on bad plant discipline and sloppy workers, and to try to police the problem away. The more flaws

---

A STAR IS BORN ON THE FACTORY FLOOR
Lisa Miller Mesdag

Armed with an MBA from Columbia, Peter Smith landed a prestigious job in General Signal's mergers and acquisitions department, where he worked closely with the chairman. But when he heard that the company was creating a post as director of manufacturing services, he volunteered and got the job. "My peers were aghast," he says.

Smith had his reasons: "There's enormous excitement in manufacturing. You transform raw materials into something of value and deal with a spectrum of people you'd normally never come across."

In 1977 Smith was appointed manufacturing vice-president of Regina, a $50-million unit that makes floor-care appliances. An erratic performer, the operation had gone through three vice-presidents in five years. Smith found the production floor grueling. "The first thing I noticed was the extraordinary pace," he says. "We worked at a run all day. Manufacturing is unlike any other discipline in the need for immediate decisions."

At age 29, Smith was General Signal's youngest vice-president and the company's first MBA assigned to manufacturing. The union wanted to test his style. "I had only been there a few weeks when a worker struck a supervisor. This was punishable by a one-day suspension or termination. I terminated her on the spot. We had a severe lack of shop-floor discipline."

Smith believes most manufacturing problems arise when management doesn't talk with the workers. "People will flood you with ideas if you let them talk." He cites his experience with a switch-assembly operation, where workers had to twist together five pairs of wires, cap them with nuts, and stuff them into a vacuum-cleaner handle. "We had quality problems with the nuts falling off, and the workers had to wear Band-Aids on their fingers. One suggested we use a switch with push-in connections instead, but the materials cost would double. I calculated how much more productivity we would need to offset the cost and asked the union to help me prove we could get it by using the new switch." The change cut the rejection rate, formerly 20 percent, to less than 1 percent. In three years at Regina, Smith reduced the cost of assembling a vacuum cleaner from $1.56 to 88 cents, generating savings of $12 million.

He continues to be attentive to workers' ideas in his current job as president of BIF, a General Signal unit that makes water-treatment equipment. Alfred Malo, 59, a lathe operator, praises Smith for laying off 150 white-collar workers. "There used to be five of them for every one of us," Malo says. "Peter got it down to three to one. I can tell the difference in the parking lot." Smith insists that production workers be well rewarded: "If they help us make money, we have to give a little bit back to them. The guys in the office can't always get *all* the gravy."

Smith hasn't done badly for himself in manufacturing. His income has more than quintupled in six years; at 33 he is the company's youngest president and is besieged with job offers. "There was no way I could have proved how important I was to the company had I stayed in finance," he says. "In manufacturing there is a clear connection between your skills and the company's profit. If the scrap rate is five percent of materials cost and you get it down to two percent, you can get a thirty percent improvement in profit without adding a dollar in sales. Nobody else in the company can do that."

that show up, the more inspectors you put on the production line, until the question gets to be: "How much quality can we afford?" The question should be: "How many mistakes can we afford?" Efforts to improve quality should be regarded not as a cost but as a way to keep costs down and increase earnings (see Chapter 13, "The Battle for Quality"). "If you do it right the first time, you save terribly large amounts of money," says George F. Kennard, director of quality and assurance in IBM's Data Processing Product Group. "Sometimes more profit is to be made by raising the level of quality than by introducing new products."

Early last year IBM started an intensive quality program, involving rigorous training for all managers and employees. Kennard rates the corporate performance "superb compared to the rest of the American scene. But we can raise our level of expectations by a factor of ten."

THE MISSING QUESTION

Quality and productivity are closely interrelated (less waste means more salable product per hours of work) and both are strongly affected by the way manufacturers use new technology. U.S. industry is just beginning to reap the full harvest of computerized innovations that could revolutionize production processes during the 1980s: Computer Aided Design (CAD), Computer Aided Manufacturing (CAM), and robotics (see Chapter 2).

By and large, American corporations have been behind the Japanese in recognizing the potentials in these developments for reducing production costs and increasing the flexibility and versatility of factory operations. Now more and more companies are joining the race to catch up. But if the new machines are regarded merely as labor-saving devices and dropped in here and there to augment existing processes and procedures, all that will be accomplished is to arouse workers' concern about losing their jobs. The companies that benefit most from these extraordinary advances in electronic intelligence will be those with the boldness and vision to introduce them as systems around which new plants may have to be built and strategies reconstructed.

Most corporate managements like to think of themselves as bold and innovative, but the true test comes when they face big investment decisions. The financial people are ready with their sharp pencils and piercing questions: How soon will the new plant pay off? What will it do to our earnings? There has to be someone at the table to ask: What will happen to our competitive position ten years from now if we don't build the plant?

When the decision-makers have no firsthand acquaintance with factory operations, their habits of thinking are skewed toward risk avoidance. They have nothing to go by but the financial figures, and the numbers do not shed much light on such intangibles as the long-term rewards that can be expected from big technological change. The "entrepreneurial spirit" is exalted in corporate theology, but it is sometimes forgotten that the risks true entrepreneurs take are based on the confidence that comes from close acquaintance with the details and fundamentals of their business.

BORN-AGAIN COMPANIES

A few large corporations have always had such confidence. Their top managements have never lost touch with the factory. Pride in product excellence and in using innovative technology are part of their competitive arsenal. They include not only pathfinders in frontier industries, like Hewlett-Packard and Texas Instruments, but also venerable enterprises such as Corning Glass and Deere & Company.

Both Corning and Deere have kept their headquarters in small cities close to some of their plants, so executives can visit informally to get a feel of the factory ambience. When Deere's Chairman William A. Hewitt walks through the farm-machinery works in Moline, Illinois, he is greeted as a personal friend. At both these companies, production people count. Deere's senior vice-presidents responsible for manufacturing are all on the board of directors and thus obliged to think of the corporation as a whole.

At Corning, four out of seven senior line vice-presidents have manufacturing backgrounds—and their ideas are heeded when important decisions are made. Robert Turissini, who started his corporate climb in a glass-tubing plant, is now president of Corning Latin America/Asia Pacific, with responsibilities for production in nine nations. Recently he was able to persuade the board of directors to authorize a new $70-million factory in South Korea to produce glass for color-television tubes; the financial executives had opposed the move.

Other companies, "born again" believers in the production cause, have been adjusting their corporate cultures to accord with the new faith. A standout example is TRW, a $5-billion producer of automotive components, electronics equipment, and space systems. It regularly sent teams of executives to imbibe Skinner's doctrine at Harvard. It also hired John Stuart Foster, Jr., an almost legendary figure in military research and development who had run the government's Livermore Laboratory, and made him vice-president for science and technology.

In June 1981 TRW sponsored a conference in Eastlake, Ohio, on "Manufacturing in the Eighties." Its purpose was, in Foster's words, "to boost interest in the importance of excellence in manufacturing." Plant managers, vice-presidents for engineering, and directors of manufacturing from all the company's units were asked to describe their plans for the decade. Representatives of four other companies—Deere, Black & Decker, Western Electric, and Tektronix—were invited to tell the TRW audience where they were five years ago, what changes they have made since then, what challenges they now face, and what they intend to do to make sure they remain preeminent in this decade. Says Foster: "We want to learn by their acts."

TROUBLE MAY TURN THEM ON

It is not enough, of course, simply to elevate the status and influence of the manufacturing operation in corporate deliberations. Some lackluster companies—in fact, whole industries—have long been dominated by production people. If they are parochial-minded technicians, then marketing and financial considerations are bound to get slighted. Accordingly, as companies reassess their manufacturing operations, they may have to upgrade the quality of their production management. They are looking for broad-gauged people, able to hold their own in corporate policy deliberations, receptive to new technology and imaginative about using it, and skilled at getting the most out of production workers (a task that will be examined in Chapter 9, "When Workers Manage Themselves").

Leading headhunter firms such as Korn/Ferry and Boyden Associates report a significant flurry of assignments to find good production managers. The supply of talent is limited, but more executives may have their enthusiasm for manufacturing kindled by the national concern about productivity and competitiveness. "When something goes wrong," says Carl Menk, president of Boyden, "there's a professional challenge."

Looking to the future, a rising number of companies are trying hard to recruit emerging MBAs for entry-level jobs in production management. (Some believe the ideal combination is an MBA with an engineering degree too.) U.S. Steel and Republic Steel have been actively recruiting at Carnegie-Mellon for the first time ever.

For their part, the graduate business schools are eager as ever to ride an emerging trend. Courses in production are being reintroduced at the prestige schools, where (except at Harvard) they died out in the 1960s, overpowered by the appeal of marketing and finance.

However, the schools face a chicken-and-egg situation familiar in new-product marketing. They sense a national need for better production managers, but they can't afford to play to empty classrooms. And whatever intellectual curiosity the students might have, they are not anxious to invest their time in courses that, judging by past experience, have little prospect to leading to top jobs. So when corporate recruiters have sought production candidates at the business schools, they have found few students who were interested. Martin K. Starr, a professor at the Columbia Business School and one of the more fervent promoters of the production management cause, describes the situation piquantly: "We are surrounded by insurmountable opportunities." At Columbia, only 2 percent of entering students say they're aiming for jobs in production (versus 12 percent twenty years ago), and more than half of them change their minds after talking to classmates.

At Starr's urging, one of the school's advisory boards, composed of two dozen corporate executives, wrestled with the problem for several months. Its recommendations, submitted in April

---

MANUFACTURING'S DISCONTENTS

How do today's executives feel about jobs in manufacturing? They're not very ardent, according to a survey Boyden Associates made recently of executives who had approached the search firm to see about changing jobs. The sampling took in 450 men and women, half working in manufacturing and half in other functional areas.

Given their choice, 31 percent of the executives now in manufacturing would switch out of it, while only 11 percent of the other group would move into manufacturing.

Asked to rank the factors that caused them to look for a new job, the manufacturing executives voted "insufficient challenge" into first place and "slow career path" into second. They also ranked the factors that were particularly dissatisfying in their present positions. The biggest gripes: (1) "personality conflict with senior management"; (2) "conflict over business goals with senior management."

Salary ranked low as a cause for discontent, but the survey revealed a significant difference between the pay patterns in manufacturing and those elsewhere. Among the executives in the latter group, 49 percent were making from $45,000 to $75,000, 23 percent were paid more than $75,000, and 4 percent more than $100,000. In the manufacturing group, 66 percent were in the $45,000 to $75,000 bracket, but only 10 percent exceeded $75,000 and none earned over $100,000.

1981, were largely aimed at increasing the appeal of Columbia's operations management curriculum by repackaging courses and giving them more enticing titles, and by enlisting the faculty in a campaign to sell students on the importance of production.

Members of the advisory board did acknowledge, however, that the decisive effort must come from business itself. Everyone may recognize the need to upgrade factory management, but the jobs are still not competitive with others the newly minted MBA can choose from. For one thing, starting salaries are still meager. Roger Breisch, a highly regarded member of the class just graduated from M.I.T.'s Sloan School, says he "had a manufacturing offer from one company that was half what my friends were being offered in consulting."

Secondly, as Breisch puts it, "corporate headquarters is a far different world from being in a plant. Manufacturing is not only psychologically isolated but also geographically isolated." The geography is bound to discourage the MBA whose spouse is also pursuing a career and is likely to find chances pretty slim in a factory town.

Finally, the ambitious MBA has no assurance that, if he takes a job in manufacturing, he won't get stuck there forever. Roger Breisch *is* taking a job in manufacturing—with American Cyanamid. His revealing reason is that "they have a program for developing managers."

The career path that corporations offer a prospective recruit can be the most powerful factor in the candidate's choice. A development program that moves people from function to function may be the best way to bring the factory back to where it belongs in the corporate fold. It ensures that production managers will get their judgment broadened by exposure to marketing and finance, and executives in those and other corporate areas will get to know what life is like in the plant. The CEOs who emerge from this kind of training would surely be a lot less prone to the kind of managerial lapses their predecessors now have cause to regret.

July 13, 1981
Research associate: Lisa Miller Mesdag

5

THE QUALITY OF WORKLIFE

CHARLES G. BURCK

It was a quiet little gathering, but momentous nonetheless. Pontiac held a dinner at Windows on the World, the restaurant atop Manhattan's World Trade Center, in May 1981, as it has several times over the past few years, to give a small group of auto writers the chance to talk informally with its top man. But this time another leader sat at the end of the table next to general manager William E. Hoglund: Ted Creason, chairman of the United Auto Workers' Pontiac shop committee. Creason and Hoglund were there to explain how important Pontiac's "quality of worklife" program has become. Without question, they said, the future of the division—and indeed of the entire industry—would be determined by such cooperative labor-management efforts. They would deliver the same message the next morning on ABC's *Good Morning America*—where substitute host Steve Bell remarked in the tones of a man impressed, "It's obvious that there's a new day in the American auto industry."

It's at least possible, over the long haul, that there's a new day for American industry in general. A powerful movement is under way to reexamine and, as necessary, break with old managerial assumptions and formulas. Today's executives are confronting the knowledge that the business system they mastered is no longer the world-beater it used to be. They are considering options—some inspired by the Japanese—that are more flexible and participative than the rigid hierarchies they grew up in. These alternative organizations take a long-run view of corporate self-interest, and are guided by a sense of common purpose that motivates all who work within them. Above all, management focuses attention on how goods and services are actually produced—a concern that should be basic and reflexive but that, in fact, has been sloughed aside to an astonishing degree by a generation of executives preoccupied with finance, marketing, and strategic redeployment, and trained to regard the art of management as an abstraction.

THE MOST POPULAR DEVICE

This chapter and later ones will examine the new managerial movement, in settings that range from the shop floor through middle management to the chairman's office—and the union hall too. The epicenter of change lies in the way people work together. No area of management is richer in potential payoff, especially now that a decade of experimentation has tested the major theories. The undertaking Hoglund and Creason discussed is part of a whole family of approaches to managing work that have come to prominence in recent years and that are at the leading edge of the movement to break with the rigidity and formalism of the past.

This family still has no generally accepted name, nor even a very clear set of definitions for the many programs, plans, and concepts within. The most recognizable and popular device is the so-called "quality circle," in which employees meet regularly to solve production problems and thrash out better ways of working. But this tactical device may be part of a wider formal undertaking called "quality of worklife," "work innovation," "work redesign," or any one of a half-dozen other names. If truly comprehensive, the undertaking may be known as "open-systems management" or "participatory management"; or it may take its name from a scheme like William G. Ouchi's "Z" management proposition or the "7-S" model promoted by Richard T. Pascale and Anthony G. Athos. The endeavor can become a major redirection of an entire management culture, such as the one Westinghouse is attempting (see Chapter 6).

By whatever name, and on whatever scale, this kind of effort is most broadly described as the process of expanding the responsibility and influence of rank-and-file employees. It assumes that people want to work together in common purpose, and it challenges the sharp distinction, inherent in classical Western industrial organization, between the actual work of producing goods or services and the planning and coordination of that work. Today's employees, it holds, are able and willing to participate more fully in management decisions at all levels, and the organization that does not let them do so not only turns them off but also wastes valuable intelligence.

The concept is most usually called "quality of worklife" (it is the name we will use here) and it has been introduced on the shop floor. But the basic principles are really an extension of the idea—almost a truism of modern management theory—that authority should be delegated to the lowest possible level. The principles are applicable not only to the factory but also to office and service workers in banks, insurance companies, government bureaucracies, and even the military. Although the concept is only now emerging as a practical managerial concern, it is likely to become a full-blown management discipline.

Like many appealing new ideas that have the ring of observable truth, the concept is loaded with dangers. Countless companies have tried work innovations and failed miserably; countless others will undoubtedly do so in years to come. Quality of worklife challenges a system of authority and accountability that has served through most of history, and people who think of it as simply another personnel program seriously underestimate it. If it isn't entirely a leap into the unknown, it nonetheless leads into a wilderness of human emotions and power relationships.

HARD-BOILED CONVERTS

Programs to improve work have been around for well over a decade, but for the most part executives have looked on them as curiosities. At best—so it seemed—the programs were laudable efforts to make employees happier, but with no clear-cut payoff for the corporation; at worst, they struck hard-boiled managers as unconscionable attempts to further coddle an already overindulged generation.

Most executives may still harbor suspicions about the programs, but the converts are becoming too numerous and important to ignore. "You can't drive a good work force thirty percent harder," Eastman Kodak chairman Walter A. Fallon told conferees at a productivity conference in 1979, "but we've found that we could often work thirty or fifty or even a hundred and fifty percent *smarter.*" Working smarter, Fallon said, means imparting a strong sense of teamwork and giving employees more say about how they do their jobs.

At AT&T a portentous memo that chairman Charles L. Brown sent in 1980 to top officers and the heads of the twenty-three Bell System operating companies made it clear that quality of worklife was to be a long-term commitment in the organization. "We are dealing with nothing less than management style here," Brown wrote. "I am speaking of an insistence on your part that the principles involved be tried out—and 'tolerated' where they cause waves—with the end objective of a gradual spread to the whole organization." General Motors has been developing quality-of-worklife programs for a full decade, and president F. James McDonald, a longtime partisan of the concept, declares unequivocally: "As far as I'm concerned, it's the only way to operate the business—there isn't any other way in today's world."

A growing number of labor leaders are speaking a similar language. A decade ago, about the only union advocate of quality-of-worklife programs was Irving Bluestone, then head of the United Auto Workers' GM unit. In the past two years, unions such as the Communications Workers of America and the United Steelworkers have joined the UAW in pursuing quality-of-worklife improvements jointly with management. Most recently, Thomas R. Donahue, the AFL-CIO's secretary-treasurer and long a scornful opponent of European-style partnerships between labor and management, declared the time had come for "a limited partnership—a marriage of convenience" with management, to work for higher productivity by such means as quality-of-worklife programs.

No area of management has been more neglected than improving the way people work together. To be sure, there have always been "people managers" with the gift of inspiring loyalty and outstanding performance (though many executives only *think* they are good people managers). And what Thomas J. Peters of McKinsey & Co. calls "obsessive attention to people in every aspect of the business" is a fundamental part of the culture at well-run companies as diverse as IBM, Delta Airlines, Hewlett-Packard, and Walt Disney Productions.

But both the people managers and the obsessively people-oriented companies are relatively rare. The typical American manager today holds forth in a rigid and stratified system that is the organizational equivalent of a multistory nineteenth-century factory building. One reason for this is because it's risky to run experiments that threaten to disrupt the established flow of work and the delicate balance of power that determines how an organization performs. Confronted with the need to change something major, most executives will step gingerly around the whole business and concentrate on more manageable risks—switching to a different ad agency, say, or selling off an operation.

RULES FOR ANY PLOWBOY

Of course, the old factory did serve its purpose well once upon a time. Classical industrial organization reflected what organizational behaviorist Gerald I. Susman calls "the rhythm and ethos of the industrial era." The machine provided the means for large-scale production, and jobs were designed to satisfy the requirements of the machine: they were broken down into the simplest, most unspecialized tasks possible, so that any plowboy fresh from the farm could, with a minimum of training, be put into a job in the mill that would not tax his ingenuity. Precise rules delineated the tasks, and a rigid hierarchy of authority ensured that the rules would be followed. Factory owners also liked the idea because

HOW WORKERS OF THE WORLD GET THEIR SAY

	Sweden	Germany	France	Britain	U.S.	Yugoslavia
Collective bargaining	★	☆	☆	★	☆	☆
	★	☆	☆	★	☆	☆
	★	★	★	★	★	☆
	★	★	★	★	★	☆
Employees help decide how work will be organized	☆	☆	☆	☆	☆	☆
	☆	☆	☆	☆	☆	☆
	★	☆	☆	☆	☆	☆
	★	★	★	☆	★	☆
Worker representatives on board or policy councils	☆	★	☆	☆	☆	★
	★	★	☆	☆	☆	★
	★	★	☆	☆	☆	★
	★	★	★	☆	☆	★
Partial or complete employee ownership	☆	☆	☆	☆	☆	★
	☆	☆	☆	☆	☆	★
	☆	☆	☆	☆	☆	★
	★	☆	☆	★	★	★

workers with low-level, unspecialized skills were cheaper and more interchangeable than specialists—and hence replaceable if they did not follow the rules.

The system was refined at the turn of the century according to Frederick W. Taylor's principles of "scientific management," which further reduced and simplified jobs in the pursuit of efficiency. (A contemporary of Taylor's, Frank B. Gilbreth, a well-known management engineer, claimed to have found the ultimate

The first wave of interest in participative management in the U.S. ten years ago got a bad press. Many people identified the movement as "industrial democracy," a European term that conjured up visions of unionists setting corporate policies or workers voting to override their supervisors. But such power sharing is only one of several ways to give employees a bigger say.

The black stars in the columns to the left show the extent to which six countries employ four different methods of sharing power. The table is adapted from work done by Eric Trist, a pioneering theorist in quality of worklife who is a professor at York University in Toronto and professor emeritus at the Wharton School. Trist's method can be used to compare worker participation not only among countries but among companies and industries as well.

Conspicuously absent is Japan; as Trist observes, Japanese management defies clear-cut categorical comparisons with the West. Japanese corporations are quite hierarchical but also exhibit a spirit of cooperation that transcends hierarchy. The combination, which mirrors the Japanese social order, narrows the gap between labor and management.

In most of Europe, where labor relations reflect a heritage of rigid class distinctions, industrial democracy serves as an essentially political mechanism for sharing power. The means vary from country to country: England is lopsidedly reliant upon collective bargaining, while Germany leads the West in codetermination. Like most other Communist nations, Yugoslavia has no collective bargaining. But its singular system of market socialism invests ownership and ultimate control of management in the workers, whose pay is linked to profits.

Labor in the U.S. has generally shown no more enthusiasm for these forms of power sharing than management. The evolving quality-of-worklife movement, however, could well become what work specialist Ted Mills calls "a uniquely American form of industrial democracy." It is driven not by ideology, but by the pragmatic notion that cooperation serves both management and labor by yielding a better product at lower cost.

irreducible time-and-motion unit for any given task; he coined a term for it—"therblig"—by reversing his name, but he never made quite the impact Taylor did.) Taylor, ironically, faced stiff opposition at first from management, which feared that his concepts would make managers just as subject to ironclad rules as workers. But he believed ardently, if naively, that workers would welcome his innovations once they realized scientific management would bring them more money for less effort.

Taylor was not entirely wrong. The highly fractionated jobs of industry could not provide the sense of purpose that informed the work they replaced—crafts, small-scale enterprise, even agriculture. But for decades the loss did not seem deeply felt by workers, whether they labored in dark satanic mills or what behavioral scientist Frederick Herzberg calls "bright satanic offices." The trade-off was an ever-rising standard of living, purchased with the ever-rising productivity made possible by industrial organization.

The first serious questions about that system were raised during the 1920s and 1930s by the celebrated Hawthorne experiments, conducted at Western Electric's Hawthorne Works in Chicago, which suggested that workers' attitudes could influence industrial output as much as, if not more than, higher pay or improved working conditions. But not until the 1950s did the fledgling discipline of behavioral science begin to produce substantial criticism of the system. In 1952 two Yale researchers, Charles R. Walker and Robert H. Guest, produced *The Man on the Assembly Line,* a study of auto workers, which showed that the fractionated jobs typical of mass production yielded high rates of dissatisfaction, as measured by turnover, absenteeism, and grievances. More broadly, the study showed—for anyone who cared to pay attention—that many workers were eager to take more responsibility for their work. Some of the workers' comments sound startlingly contemporary today. "One of the main things wrong with this job is that there's no chance to use my brain," said one. Another declared poignantly, "All I do is think about all the things that went through wrong that should have been fixed."

A HIGHER ORDER OF NEEDS

Most of the major concepts that underlie the quality-of-worklife idea were developed in the U.S. during the 1950s and 1960s. The humanistic psychologist Abraham Maslow, for example, pinpointed the limits of the classical industrial organization's capacity to satisfy its workers; as each basic human need was met, he observed, people would develop new and higher orders of needs, and material rewards would have to be supplemented with psychic gratifications.

Douglas McGregor, a professor of human and industrial engineering at M.I.T., followed with a strong humanist critique. The management style of the typical rigid and autocratic organization, he declared, reflected a belief—"Theory X"—that people inherently dislike work and want to avoid responsibility. He advanced Theory Y as the opposing paradigm of enlightenment: work is as natural as play or rest, people want responsibility, and "the limits on human collaboration in the organizational setting are not limits of human nature but of management's ingenuity."

Rensis Likert, professor of psychology and sociology at the University of Michigan, interpolated between McGregor's concepts a spectrum of organizational types, ranging from what he called System I—the most rigid and hierarchical—to System IV, the most participative, cooperative, and flexible. Chris Argyris, professor of organizational behavior at the Harvard School of Business, analyzed vigorously the conflict between the individual and the organization to show how the classical system, by its very nature, set management and employees at cross-purposes and sabotaged any efforts to bring them together to work more effectively.

For the most part, social scientists in the U.S. who studied the organization and the worker came, like McGregor, from the humanist tradition, and focused their efforts on the individual job or on relationships among individuals. A more comprehensive approach was taken by a group of British social scientists at the Tavistock Institute, a social and industrial research organization, who produced a critique with the awe-inspiring name of "sociotechnical systems theory." They held out little hope for significantly changing an industrial organization without taking into account the demands of its technology. The combination of social and technical systems that made up any given company were interrelated; even if all other things were equal, the dictates of technology would make working relationships in a steel mill different from those in an auto-assembly plant (to say nothing of an electronics plant). The success of the organization would depend on what Eric Trist, one of the school's leading lights, called "the goodness of fit" between the social and technical systems—a fit that would be different for every company.

The socio-technocrats' rather sophisticated theory anticipated the broader approach to developing quality-of-worklife programs that prevails today, but it did not make much headway in the U.S. for many years. Indeed, even the stronger and more eloquent humanists were unable to make a case against a system that seemed to be steaming along splendidly, and in which the rewards of work still outweighed dissatisfactions in the workplace.

The cultural upheaval of the late 1960s and early 1970s brought a flurry of interest in work reform. Even as American businessmen scratched their heads over the youth rebellion, they were discovering the related problem of the alienated worker and the "blue-collar blues." Social scientists and fugitive personnel-staff specialists of all sorts came forth, eager to prove their theories about remedying the deficiencies of the Taylorized job—or at least to make a buck.

Some of the results were good. A number of today's most important quality-of-worklife programs began back then—Procter & Gamble's started in the late 1960s, Dana Corporation's in 1969, and General Motors' in 1971. More often, though, the results were dismal. In scores of companies and plants hopes were aroused only to be dashed. Some programs set up as controlled experiments did nothing more than turn the control and experimental groups against each other. Others were hit-and-run exercises by consultants pandering to the imperishable managerial fantasy of a patented quick-fix solution to complex problems. Even the best social scientists were divided over principles and tactics. Executives found it easy to ignore what was going on: "job enrichment" had the ring of hortatory bilge and "industrial democracy" of leftist eyewash.

The current wave of interest in the quality-of-worklife concept is something else again. It has come on so suddenly that it has caught many longtime partisans off-guard; some wonder dubiously if it is not a passing fad. But the climate now is strikingly different from that of the past. Today's corporate leader is often as not reviewing with dismay his stagnant productivity and substandard quality, while glancing nervously across the Pacific for salvos from the competition; he is apt to be relatively open-minded about changing his organization.

The dissatisfactions of industrial—or clerical—work are no longer disputed. Even in the 1950s, as Walker and Guest revealed, workers wanted to do more than the system let them do. Since then, the exponential growth of information in modern society has raised the expectations and sophistication of even its least educated members.

Every significant poll or survey further underscores the likelihood of a payoff. A major survey of U.S. workers' attitudes toward productivity conducted by the Gallup Poll for the U.S. Chamber of Commerce in 1979 found "the overwhelming majority believe that if they are more involved in making decisions that affect their jobs, they would work harder and do better." A more recent international survey on productivity, developed by pollster Louis Harris and sociologist Amitai Etzioni and released in April 1981, echoed the point in some of its findings, and produced a striking complementary message about the kind of human resources that lie untapped. Asked what sacrifices they would make to "increase investment to achieve economic growth," nearly three-quarters of the employees surveyed said they were willing to be assigned to work wherever they were needed in their company, and nearly two-thirds would be happy to have their salaries linked to higher productivity.

CHAUTAUQUAS OF A NEW WORK ETHIC

Perhaps most important, executives are hearing more and more about what innovative approaches have accomplished in the real world. Information gushes from such institutions as the Work in America Institute in Scarsdale, New York. Forums, meetings, and informal networks of interested businessmen abound, like chautauquas of a new work ethic, where executives receive inspiration and trade experiences. The theoretical sniping that used to divide the social scientists has been submerged in a growing body of practical knowledge and experience. Says Ted Mills, chairman of the American Center for the Quality of Work Life, "In places where it's working, like GM, the ownership of the idea has passed from the organizational development people to the line managers."

Yet the skeptics have a valid concern. A great many managers are sure to latch onto one type of program or another—say, quality circles—for a fast fix of a floundering operation; these people will be lucky to escape without doing their organizations serious injury. Others will try to impose a quality-of-worklife program by fiat, with similar dubious prospects. What is perhaps the best definition of quality of worklife yet comes from Charles Bisanz, a consultant to the Northwestern Bell Company: "It is not a solution devised by management so much as it is a decision by management to share responsibility with employees in devising solutions." The distinction is likely to be overlooked by Theory X managers.

The greatest potential for trouble lies in middle management. "I find it's easy to get top executives on board," says Rosabeth Moss Kanter, a sociologist and a consultant in the field. "They like to be intellectual, they're impressed by professors, they're intrigued by concepts, they tend to be moved by examples of how other companies are doing it. They say, 'This is wonderful, let's do it a level below.' But to me that group down below is much more difficult. They'll have to pay the price."

Middle managers have spent their working lives polishing the skills required for survival in hierarchical organizations. The principles inherent in the quality-of-worklife concept can make their jobs better over the long run—but not until they've acquired a new set of skills for dealing collaboratively with the people who work for them.

Any executive contemplating a quality-of-worklife effort should be prepared for the possibility of fundamental, long-term changes in the entire organization—changes in traditional internal relationships and lines of authority. And, of course, the process will take time: GM, for example, has been at it for a decade and is still barely halfway toward getting quality-of-worklife programs into all its plants.

In the end, relatively few managers may be able to take the risks involved—or handle them. Yet it is hard to escape the sense that the principles underlying the many concepts of working smarter are right for now. Certainly it seems far less likely that the system of industrial organization perfected at the end of the nineteenth century is still the best one for the end of the twentieth.

June 15, 1981
Research associate: Faye Rice

6

WESTINGHOUSE'S CULTURAL REVOLUTION

JEREMY MAIN

At Westinghouse Electric Corporation something strange is going on: a sizable part of the company is converting to Japanese-style management. Westinghouse hopes to achieve dramatic improvements in productivity by trying a form of Theory Z, described by William G. Ouchi in a new handbook for American businessmen who want to follow the Japanese way. The effort is producing a cultural revolution at Westinghouse, overturning old-style boss-employee relationships.

The company's construction group, which represents 7 percent of its work force, offered itself up as guinea pig for the experiment in 1980. The new method rests on the theory that if labor and management work at achieving a Japanese-style consensus, Westinghouse will get better ideas, better decisions, and better execution. So today, in the black steel corporate headquarters in Pittsburgh's Gateway Center, bosses in the construction group don't simply issue orders; they seek consensus. Out in the factories, foremen don't bellow commands coarsely at workers—at least they aren't supposed to; they ask for suggestions instead. Everywhere new committees and councils are meeting on office time to discuss matters as nebulous as group synergy and as critical as next year's capital allocations.

Theory Z is a kind of "participative management," which is

hardly novel (see the preceding chapter). Social scientists have been advocating participative management for years, and many American companies have tried it, up to a point. But to see it seep into a hierarchical old industrial company like Westinghouse, with its established chain of command and staff of tradition-minded engineers, is a bit like watching the U.S. Marines parade in blue jeans, long-haired and unshaven.

The experiment began as a drive to increase productivity. Westinghouse had gone through tough times in the 1970s: it made a series of bad acquisitions, got involved in a nasty bribery case, and lost a pile on consumer appliances before selling the business. Worst of all, Westinghouse agreed to supply uranium under fixed-price contracts—an appallingly risky decision that ultimately will cost the company nearly $1 billion. To recover lost ground and meet the escalating challenge of the Japanese, the company decided two years ago that it would have to increase productivity much faster than the 2 or 3 percent that U.S. industry achieves in a good year.

Participative management might seem a roundabout route to productivity; it certainly isn't a quick fix. Westinghouse expects to wait two years before seeing any results, ten years before the benefits take full effect. The corporation isn't betting all its marbles on participation: it's also testing a panoply of other devices, ranging from a form of "matrix" management with changing and flexible lines of command to heavy investment in automation. Some industry and academic critics argue that the degree of worker participation is irrelevant to productivity; they say that workers respond to challenge, responsibility, advancement, financial rewards—not simply to being asked to participate in decisions. Skepticism about participative management exists at Westinghouse, but the highest-ranking official close to the productivity drive, vice-chairman Douglas D. Danforth, says he supports the idea warmly.

In the construction group they talk of participative management with the ardor of those who have seen the light. "The point is, we are making much better decisions than before," says Donald W. Neukranz, who runs the group's elevator division in New

Jersey. "We are getting a contribution and commitment from larger numbers of people. The management team becomes excited and it works." When participative management catches on, according to the faithful, decisions are carried out by enthusiasts who have helped shape them, who feel they "own" the decisions, rather than by unwilling subordinates who have simply been told what to do without really knowing why.

EVERYONE A TIGER

Westinghouse executives think participation is a secret of Japanese success. "When you visit Japanese factories and see everyone, but *everyone,* working like tigers to make that product more reliable at a lower cost, it's awesome," says William A. Coates, 52, the executive vice-president who runs the construction group. "They even come back early from their breaks. In factory after factory, everyone inside is trying to whip us. If we don't get that attitude, we literally won't survive."

The Westinghouse management council, which annually convenes the company's 225 or so senior executives to discuss solemn undertakings, recognized the importance of productivity when it met for two days at the Tamarron resort hotel in Durango, Colorado, in 1979. Vice-chairman Danforth appointed Thomas J. Murrin, 52, president of the Public Systems Company in Westinghouse, to head an ad hoc committee and gave him $20 million to explore ways of increasing productivity. As Murrin describes it, Westinghouse sort of backed into the policy: "Our operating margins didn't look as good as we hoped for the future, and we agonized a lot over this. The significant and delightful development came when we freed ourselves from trying to solve the problem by changing the mix or getting the volume up or raising prices. We said, realistically, these things are not fully, and sometimes not at all, under our control. Maybe we had better concentrate on things we can influence. We are going to have to do more with less—fewer people, less money, less time, less space, fewer resources in general—and I think that's probably a pretty good definition of productivity."

Murrin had a predisposition for participative management that you might not expect to find in a man so blunt and burly. He remembers that when he was growing up on New York's East Side his father, a structural steelworker, used to explain to him what a "dumb ass" his foreman was. It taught Murrin the ordinary worker could contribute a lot more to his work than muscle. Playing tackle for one of the Fordham University squads coached by Vince Lombardi added another element to Murrin's philosophy of productivity: he learned the power of teamwork.

Murrin carries a lot of weight at Westinghouse. The corporation is divided into four major companies—International, Power Systems, Industry Products, and Public Systems. As president of Public Systems, Murrin turned an unpromising hodgepodge of defense electronics, soft-drink bottling, real estate, and other operations into the fastest-growing company of the four. Public Systems in turn is divided into four groups. The one experimenting with participative management—the construction group—makes equipment for the construction industry—elevators, office systems, fans, heating and cooling equipment—as well as rapid-transit equipment for cities and people-movers for airports.

When Bill Coates was promoted from president of the elevator company in 1979 to run the whole construction group, he had no particular convictions about productivity. But he is the sort of fellow who gets up at 4:45 a.m. to run five miles and reaches the office by 7:00 to 7:30. He can develop enthusiasms as strong as Murrin's. Setting out to discover what others were doing to improve productivity, Coates and his boss visited Japan and sent teams of specialists there. The more they learned, the more they became convinced the Japanese were showing them the way.

They were especially impressed by Bill Ouchi, 37, the author of *Theory Z* and a professor of management at the University of California at Los Angeles. Ouchi produced a videocassette explaining how the Japanese achieved growth rates rarely matched in the West. Murrin and Coates saw the cassette and invited Ouchi to Pittsburgh to explain Theory Z in August 1980.

They bought participative management in general but by no means the whole Japanese system. Indeed, it would not only be silly but illegal to imitate the Japanese in every way. For instance, Japanese women are systematically excluded from management. In addition, the Japanese believe in evaluating employees infrequently and promoting them slowly, policies that would send the best and brightest at Westinghouse streaming for the exits. Nor is the construction group about to offer lifetime employment to its workers, though it is edging toward a policy of fewer layoffs. When one of the construction group's units suffered a huge drop in orders in February 1980, the company did not send workers packing, as it would have before. "We want to think of employees as family," says Coates. "You don't lay your brother off."

Westinghouse made some mistakes as the participative-management drive got under way. Ouchi was concerned that the construction group was starting the process at the wrong end by launching it on the factory floor. Until executives start making their own decisions by consensus, he says, efforts to install participative management down in the ranks will almost certainly founder. Coates and Murrin saw the point. They also knew that managers and white-collar workers represent half the corporation's work force and 70 percent of its payroll, so greater opportunities for increasing productivity could be found in the office than on the factory floor.

Coates discovered that building participative-management teams and defining their roles required subtle leadership—and a degree of hypocrisy. You can't rely on spontaneous forces to set up quality circles, managers' councils, and the like, yet if the boss simply orders participative-management groups established, the process of setting them up wouldn't be truly participative. A similar dilemma arises when new participative-management teams start trying to make decisions. Should the boss watch as the group flounders in disagreement? Should he let stand a decision reached by participation if he knows it to be wrong? Or should he intervene and undermine the participative ideal? "Sometimes the boss has to nudge his people in a nonauthoritarian way," says Ouchi, who agrees that a little hypocrisy can sometimes be effective.

Coates reserves the right to make unilateral decisions or overrule a consensus, but he must use his power sparingly if participation is to take root. He is gradually submitting more and more decisions to the consensual process. The transition, he admits, "is very tricky, very difficult." But as things have turned out, he says, he has been surprised at the soundness of decisions reached and has vetoed only two or three.

To get the participative process rolling, Murrin and Coates relied on a classic ploy management uses when it needs support for what has already been decided: they called in consultants. Dozens of consultants gave seminars and lectures, led team-building and sensitivity-training exercises, and ran courses for future team leaders or "facilitators," as they are called. Coates further prepared the ground for seeding by recruiting members for his staff who had shown enthusiasm for participative management in other parts of his group.

In this setting, participative committees, councils, and circles began sprouting in the fall of 1980. Ouchi became chairman of an outside committee of three consulting academics; Coates's staff formed three quality circles and a business strategy board. The general managers of the construction group's five units, who had rarely met to discuss issues of common interest, created a council that meets monthly. They also set up ten councils under them, in which controllers, personnel directors, marketing directors, and other specialists from each of the units discuss common problems with their counterparts in other units. In the plants, which already had limited experience with participative management, workers and supervisors formed sixty quality circles. (Westinghouse as a whole has adopted the quality-circle idea and now has more than six hundred circles, with three being added every day.)

"PRIORITIZING THE AUDIBLES . . ."

To an outsider, the functions and achievements of these new groups are not always clear. A quality circle can wander free-form among subjects without having fixed goals. The circle made up of the group staff may have made an important contribution to the jargon of the age. They have discussed the problem of "audibles," these being defined as unexpected interruptions such as phone messages that break into the day's scheduled business. The purpose of the discussion, as one participant put it, was to figure out "how you can control your audibles so you can prioritize them."

The staff circle had been discussing the nebulous subject of group synergy for five months. At a recent meeting some of the members seemed adrift, still searching for something solid to hold onto. When Robert J. Tubbs, group legal counsel, suggested a vote on an order of priorities, he was hooted down. In the proper participative way of doing things, you don't take votes or try to impose your will on anyone—you keep talking until you reach a consensus. Chastened, Tubbs blushed and said, "Let's consensutize." The meeting ended with a fifteen-minute discussion about when to hold another meeting; partly because the members of the circle had so many meetings to go to, it was difficult to find a time to suit everyone.

By contrast, the quality-of-worklife committee, composed of workers and supervisors at Grand Rapids, Michigan, where Westinghouse makes office systems, couldn't be more down to earth. A subcommittee took responsibility for establishing an attractive cafeteria to replace a shabby vending-machine area. In the proper participative manner, management did not limit how much could be spent and accepted the subcommittee's plans, which cost $500,000 to implement. Another subcommittee tackled vandalism in the restrooms, where some workers were covering the walls with graffiti and wrecking the drinking fountains. The company fixed up the restrooms, and the union asked its members to report vandals to their steward. "People are treating the restrooms better," says Lee Raterink, president of the carpenters' and joiners' local. "It worked because we worked together, because people participated in the decision."

The council of managers from the construction group's five units has produced a string of policy decisions that would undoubtedly have come out differently before. For instance, each unit used to design its own business systems, such as inventory and manufacturing controls. The council decided that the job should be done by the group as a whole. Coates says the savings will amount to "tens and tens of millions of dollars." The council persuaded Coates's staff to reorganize so that it could focus on strategic matters rather than the day-to-day business of the plants. It also created a new program for choosing and training technical personnel, rejected two potential acquisitions, and set up a cooperative system for handling bids on contracts that involve more than one unit.

Perhaps the most important decision by the managers' council allocated capital among the units. In the past, the managers would never have met to discuss allocations. Coates or his predecessors would have received requests for funds, assumed they were inflated, lopped a bit off each, and told the managers what they were getting. This time, after listening to each other's problems and prospects in the council meeting, the managers abandoned the normal stance of each one defending his own turf. They decided certain units should be pushed hard and others cut back—in one case almost to zero. Some of the managers voluntarily gave up allocations that Coates thinks they would have fought for had he tried to make the cuts. "They helped me do a fantastic job that I could never have done myself," he says. "If I had ordered them to do what they themselves decided to do, I would have had an insurrection on my hands."

Some managers, particularly at the middle and lower levels, don't like the consensus system, and a few had to be reassigned to jobs in the group where they don't have to implement the new plan. But most have become enthusiasts, some to the point of growing tiresome on the subject, like reformed drinkers. They don't feel the loss of power many managers fear at first. "I don't perceive participative management as giving up controls," says Coates, "but rather gaining a ton-and-a-half of help. The old way wasn't good enough. Industrial systems have become much too complicated for the know-it-all manager to know it all anymore."

Blue-collar reaction ranges from cool to warm. The Grand Rapids plant reports big drops in grievances and absenteeism. It sent employees to visit customers and see how the product was performing, and they came back with suggestions. They figured out ways of attaching fabric more smoothly to the office partitions they build and improving the alignment of the fixtures on them. Last year 350 employees, about one-third of the work force, went to the office-furniture show in Chicago to see what they were up against from the opposition.

EXPLETIVE DELETED

Even at the stone and concrete Sturtevant division factory in Hyde Park, Massachusetts, a turn-of-the-century relic where Westinghouse builds huge industrial fans, old adversary relationships between labor and management are beginning to crack. The union has agreed to go along with participative management, provided it is introduced by a consultant approved by the local. Some authoritarian supervisors already seem to be melting into the participative mood, albeit gradually. Referring to one of them, a machinists' union official says, "Instead of calling me a f——g ass, now he just calls me an ass."

Although even the most enthusiastic converts agree that participative-management meetings eat up hours, Coates says lost time is recovered later. "We spend a lot of time trying to get a consensus, but once you get it, the implementation is instantaneous. We don't have to fight any negative feelings." Moreover, Ouchi predicts, once the Westinghouse people get used to the new style, meetings will go faster.

The participative system has revolutionized the role of the secretary. The group has installed communications and information systems that allow executives to record messages to one another rather than play telephone tag, missing connections because first one and then the other is tied up. When an executive has a convenient moment, he can now dial a message center and hear all his callers explain what they want. Then he can dictate a reply or forward the message to someone else. The system relieves secretaries of the endless task of taking messages. It has proved so successful that it will soon be extended from 142 users in the construction group to 1500 executives across the whole corporation.

A new word-processing center has taken over dictation and typing chores. Bosses can dictate letters and memos to the center at any hour, seven days a week, from anywhere in the world. If they want to edit the material, they can have it displayed on terminals in their offices or at home. At first, Coates says, secretaries felt threatened by the change, thinking that the center was taking work away from them. But when they saw that there were more interesting and productive things to do, they quit worrying. Although the construction group hasn't found a way of measuring office-worker productivity, Coates says he believes "production is up substantially at headquarters, and the principals have more time to work on substance." Secretaries have become administrative assistants, taking over tasks their bosses used to handle, such as organizing conferences, sitting on task forces, and preparing and presenting research data. One sign of how much the secretaries do now: when Coates checked in with his secretary while he was traveling recently, she told him she had received thirteen calls to his one—and she took care of that one too.

Given the opportunity to be creative, the secretaries have come up with good ideas. For instance, they realized they were wasting time, money, and stationery preparing separate envelopes all going to one division, such as Grand Rapids. Now they send just one large envelope a day. When the construction group staff was moving to new offices, the secretaries decided that half the files could be thrown out and that one central file could serve the whole staff. This and other improvements will reduce by one-fifth the space group headquarters occupies.

WAITING FOR THE TAKEOFF

By amassing many such small, common-sense changes, Coates and Murrin hope to show they can accelerate an improvement in productivity at Westinghouse that already seems under way. Westinghouse measures productivity by subtracting the cost of goods and services from total sales to get value added and then dividing by the size of the work force. By this measure, corporate productivity gained 2 or 3 percent a year during most of the 1970s and met the new goal of a 6.1 percent annual increase in 1979 and 1980. The Public Systems Company, including Coates's group, did even better—up 8 percent in 1980. The most-quoted government statistics measure productivity differently, by dividing output by man-hours of labor. But when government statisticians use the Westinghouse yardstick, U.S. productivity in all manufacturing shows an increase of 0.4 percent in 1979 and a drop of 1.4 percent in 1980.

Coates and Murrin are convinced their system will work. But it will take time and continuing signs of progress to win over supervisors and middle managers who continue to harbor doubts about the system. "They've seen a lot of programs come and go," says Ouchi. "They wait to see if you give them commitment. Once they see this is for real, then, whoosht, productivity takes off." Even then, Ouchi believes, the whole corporation will have to adopt the system if it is to yield the bounteous harvest it is capable of producing. "Participative management," he says, "cannot survive in an alien corporate culture."

June 15, 1981
Research associate: Anna Cifelli

7

BATTLING YOUR OWN BUREAUCRACY

JEREMY MAIN

When an engineer at Intel Corporation in Santa Clara wanted a $2.79 mechanical pencil, processing the order used to require 12 pieces of paper and 95 administrative steps. Intel is enormously successful, a developer and leading manufacturer of the tiny microprocessers that make modern computers possible. But if ordering a pencil was that complicated, Intel was not as good at running itself as it was at making microprocessors. Administrative procedures and costs were getting out of hand.

Intel recognized the danger in 1979 and began an attack on its own bureaucracy that makes it possible today for the company to get those pencils with only one form and eight steps. Intel also did the following:

- Reduced from 364 to 250 the number of steps it takes to hire a new employee.
- Cut by 29,700 a month the number of Xerox copies made in the accounts-payable department in Santa Clara.
- Started putting expense accounts through in days rather than four weeks.

The benefits of these and other reforms—most of them simple changes in office procedures—are just beginning to be felt. Intel figures that the permanent savings already amount to at least $2.5 million a year, mostly by the elimination of 153 jobs. Joseph P. Nevin, the cocky, talkative 34-year-old who has been running Intel's productivity program, explains what the reforms could mean if extended to all operations of the company. "We have evidence we can get 30 percent gains in productivity," he says. "That could be worth $60 million a year to us. And that would be like a $277 million boost in sales, assuming a 22 percent pretax profit margin [Intel's five-year average]. It would raise our after-tax earnings by 73 cents a share, and it would hardly cost a thing." Intel earned $2.21 per share in 1980 on sales of $855 million.

Intel's technique for simplifying jobs is so obvious and elementary that it's a wonder everybody hasn't always used it. No smashing breakthroughs or brilliant innovations lighted the way. The company takes each administrative procedure, examines it methodically, lays it out in meticulous detail, strips off the unnecessary work, and puts it back together in a simpler, more rational manner. A form is dropped here, an authorization there; photocopies are eliminated, files thrown out, a superfluous audit removed. Soon hundreds of little steps add up to a gigantic stride.

The process hasn't been painless. Most employees seem pleased, but others have quit or found the new methods frustrating. So far only about one-fifth of the corporation's administrative staff has been exposed to reform—and the full effects have not been felt even by them. Intel started with what Nevin calls "the low-hanging fruit," departments with routine activities, such as accounts-payable and personnel records, that are relatively easy to streamline and that provide rapid proof of the program's effectiveness. More complex activities, such as sales and engineering, haven't yet been touched.

To lessen the threat of change, Intel promised not to fire any permanent employee whose job was eliminated. The company's phenomenal sales growth, 29.3 percent in 1980, helps absorb everyone who wants to stay. Intel also reassured its employees that more productivity does not mean working harder. On the contrary, "working smarter"—the company slogan for the program—often improves jobs by eliminating the dreariest tasks.

For a company that sells microprocessor chips to makers of "smart" office machines, Intel puts surprisingly little faith in its clients' equipment as the secret to raising office productivity. "Why buy a new word processor that will turn out five hundred letters an hour when your mission isn't to turn out more letters?" Nevin asks. "You may automate a function that shouldn't exist in the first place."

SOMETHING ABOUT THE SMELL

For a long time, with sales booming and profit margins fat, Intel didn't worry about office productivity, and procedures grew haphazardly. But eventually Intel had to worry. John A. Calhoun, 37, a nine-year Intel veteran who recently moved up from corporate controller to director of business development, recalls that in 1978, "bills weren't getting paid, customers were complaining. We solved problems by brute force, by pouring in more people. Fifty percent of our people had been with the company less than six months."

Administrative costs became the fastest-growing part of Intel's activities. By 1979 half the company's 10,000 U.S. workers were in administration. Adding engineers and salesmen, nonproduction workers made up 64 percent of the total. Then one day the president, Andrew S. Grove, said, "Hey, this just doesn't smell right. We've got to do something about it."

People at Intel favor the laid-back interjection "Hey"; they also dress informally and use only first names. Known to everyone around the shop as Andy, Grove sports a curly beard and likes to work in a shirt unbuttoned to reveal a gold chain. But the casual air is deceptive: Intel didn't get where it is by being mellow. A former executive recalls that its management reviews are run with an unsparing search for truth that bruises the performer with verbal brutality, even ridicule—not out of sadism but to test his case. As this man puts it: "There are no sissies at Intel."

Grove's observation launched a high-priority effort to do something about office productivity. Laurence R. Hootnick, 39, a senior vice-president and eight-year Intel veteran, took overall charge, and John Calhoun set up a study committee in the spring of 1979. Like the vast majority of U.S. executives, Calhoun knew nothing about productivity and recalls that the committee floundered for a while, brainstorming to little effect and heading up blind alleys.

THEY STUDY THE MOTION

In their first firm step forward, Hootnick and Calhoun called in Wofac Company, a New Jersey consulting firm that specializes in a system for improving productivity marketed under the name VeFac. The process is a variation of the old time-and-motion studies devised at the turn of the century by Frederick W. Taylor. A Wofac consultant studies a unit such as the supply department for several weeks, determines how often each job is performed, and how long it should take, and then estimates how many workers the unit needs—usually fewer than it has.

Wofac attacked several areas of Intel, mainly inventory and supply functions, and quickly produced results. David J. Hamilton, one of Nevin's deputies, says that Wofac had already saved Intel its $240,000 fee by the time the fee was paid. Following the VeFac plan, the company made fixes that eliminated 122 jobs and now save $1.9 million a year.

But VeFac is a canned, standard system that has its limits. Supervisors were sometimes infuriated by it. One complained: "That [Wofac] guy comes in here and takes notes for a couple of days and he comes back a week later and says, 'You need five less people.' What the hell does he know?"

Even while taking on Wofac, Calhoun and Hootnick sensed they needed a broader, custom-made approach to productivity, and that led them to Joe Nevin. A Californian by enthusiastic transplantation, Nevin has the air of a schoolboy who has just outsmarted his teacher. He was graduated from the University of Southern California, then earned an M.S. degree at M.I.T.'s Sloan School of Management. He went to work for Citibank in Manhattan in 1972, on the staff of Robert B. White, now an executive vice-president, who was trying to rescue the bank's backroom operations from monumental disaster. Citibank had grown rapidly in the 1960s and tried to cope with the inundation of paperwork by buying huge computers and adding thousands of clerks. By 1970 the backroom employed 10,000 people, and customers raged about delays, errors, and their helplessness in getting mistakes corrected. Backroom costs were climbing 15 percent a year.

White decided that the system was the problem. The 10,000 employees were not lazy or incompetent but badly managed. They were scattered along a huge pipeline that passed each transaction through many hands. The backroom was organized by function: all check encoding was done in one place, all computer entries in another. Responsibility was lost; errors couldn't be corrected. White first simplified the work, eliminating all the superfluous steps he could find. Then he completely reorganized the backroom, replacing the functional divisions with small work stations equipped with small computers. Each group was responsible for a particular type of transaction, such as processing a transfer of funds from abroad, and one person often performed all the steps required. Errors could be found quickly; responsibility was clear. Service improved, the backroom staff was cut by a resounding 40 percent—and Nevin learned how to bring a business bureaucracy under control.

After six years at Citibank, commuting three hours a day between Manhattan and New Providence, New Jersey, Nevin was sipping mulled wine with his wife one winter's night when an ice storm broke the power lines and left them in the dark. He turned to his wife and said, "By September, we're going to be in the Bay Area." And on September 14, 1978, he went to work in San Francisco for the Bank of America, which wanted his know-how.

The world-banking division where Nevin worked was too tangled in internal politics to let him do much about productivity. So when Intel asked him to be "czar of productivity," as he puts it, he accepted. Besides, he says, he had been frowned on at the bank for removing his jacket, as he had been at Citibank for wearing a beard. (He has since shaved off the beard now that no one minds it. At Intel he could probably come to work in beads and earrings without breaking any canons of corporate behavior.)

Nevin's efforts at the Bank of America weren't altogether wasted, according to a source in the bank, because after he left, the measures he had tried and failed to implant in world banking were adopted elsewhere in the company.

MEASURING THE JANITOR

Nevin begins a job of work simplification, as he describes his system, by defining and measuring the productivity of an office. He calls this "getting rid of the smoke." It's not as easy as it might seem. "People often can't distinguish between activity and output," says Nevin. "When I asked people at the Bank of America what they did, they talked about visiting plants, going to conventions, lunching with clients, reading *The Wall Street Journal,* and so forth, but they had lost sight of what they really did. We decided that was establishing lines of credit and making loans."

Managers and office workers resent the outsider who wants to gauge their output. Almost invariably, Nevin reports, they say, "Hey, you can't measure me." But he claims he can. He hasn't yet found an office where he couldn't apply some rudimentary but effective measure: How many vouchers did accounts-payable issue? How many people did the employment office hire? How many square feet did the janitors clean?

Nevin is a fanatic about defining productivity, which he feels most businessmen misunderstand. He rejects any definition that involves dollar values. Consideration of dollar values, he argues, belongs with the separate and equally important question of profitablity. He says productivity should be measured by dividing physical output in units by man-hours of work. For example, the output of the accounts-payable office is found by dividing the number of payment vouchers processed by the hours worked.

To keep track of progress in a department, Nevin sets up a productivity index that is based on output just before the simplification begins. In the case of accounts-payable, the base was the second quarter of 1980. In those three months, the department's seventy-one employees processed an average of 33,000 vouchers per month, or one every twenty-three minutes. Accounts-payable set a productivity goal of sixteen minutes per voucher.

Without quality standards, a productivity index could be misleading; accounts-payable might start working faster but churning out vouchers full of errors. So the system requires a quality index alongside the productivity index. The accounts-payable department checks quality in a variety of ways. For instance, late payments to vendors must not exceed 2 percent of the total.

Once the rudimentary productivity and quality measures have been designed and "there's no smoke," as Nevin puts it, the painstaking task of simplifying begins. Intel believes in participative management, so the people whose jobs are to be simplified are asked to do it. The office under scrutiny sets up a group consisting of its chief, two or three other managers, and two or three clerks. Nevin or his assistant for job simplification, Bette J. Lloyd, attends the group's meetings to encourage, advise, and prod.

Over several weeks, the group lays out in minute detail every step in a particular process on huge charts similar to the one depicted here. The more complicated procedures cover three walls of a conference room. The steps are represented by symbols: a triangle for filing papers, a square for a checkpoint, and so on. Each step—even one as minor as "walk to files"—is examined to see if it can be cut. Nevin's favorite question, often put in the mischievous manner of someone who thinks he's going to get a silly answer, is: "Hey, why do you *do* this?" Those who give a wrong answer, such as, "We've always done it this way," are awarded the appropriate "working dumber" ticket from a book supplied by Nevin. As the meetings progress, suggestions for improvements are accumulated on a list beside the chart.

When the accounts-payable procedures were being simplified, the chart showed that freight invoices were checked both by Intel's auditors and Northwest Traffic Associates, Inc., a company that specializes in verifying freight bills. Northwest was hired

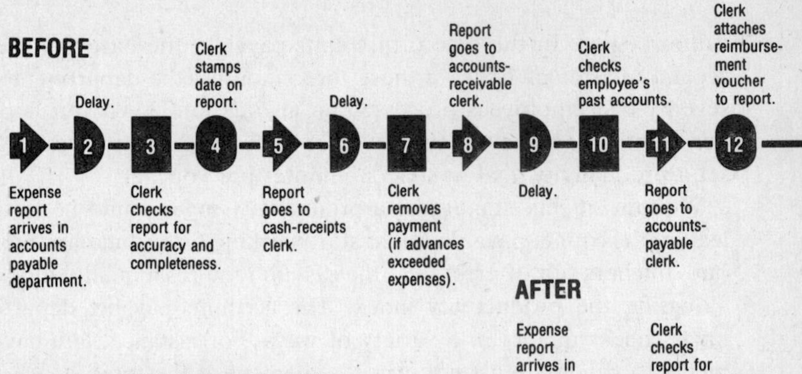

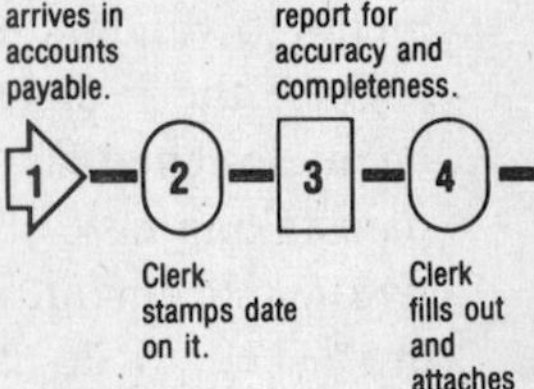

THE EXPENSE-ACCOUNT EXPRESS

Intel simplifies work by charting the steps it takes to do something and then removing as many of those steps as possible. In this before-and-after example, the handling of expense accounts was reduced from 25 steps to 14. The accounts-payable clerk took over the cash-receipts clerk's job of collecting refunds or unused traveler's checks, eliminating steps 5 and 7. The accounts-receivable clerk's job of checking the employee's past expense accounts (steps 8, 10, and 11) was eliminated; another department already did that. Checking items against company guidelines (14) was judged more trouble than it was worth. Logging batches (19) proved unnecessary. Four delays (2, 6, 9, and 18) were cut along the way. Expense accounts are now processed in days rather than weeks.

---

before Intel had developed its own checking capacity, and no one had thought to amend the contract. The simplification group decided to drop most Northwest audits and saved $1475 a month.

At another meeting, the discussion turned to expense accounts. Why should an employee complete a full expense report and wait a month for reimbursement if he is only trying to collect for a business lunch? The group decided that for expenses under $100, the employee should just fill out a petty-cash voucher and collect immediately from the cashier. Intel hasn't gone so far as to pay outside suppliers that fast; they still wait fifty-six days. Because the company gets interest-free use of the money it owes suppliers, it doesn't carry efficiency too far.

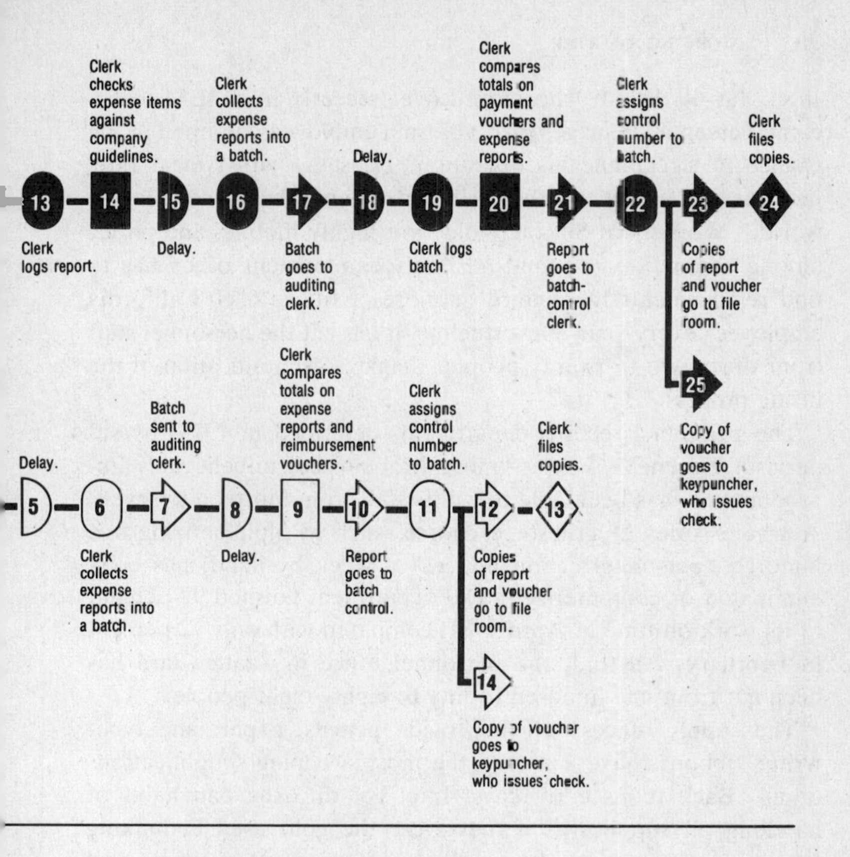

After simplifying more than a dozen other types of transactions, the accounts-payable department cut its staff from seventy-one in 1980 to fifty-one in April 1981. And it achieved its sixteen-minutes-a-voucher goal.

If he were simplifying accounts-payable all over again, John Calhoun says, he would go slower and make managers pay more attention to the personal stress and frustration created by new routines and other changes. Many employees were enthusiastic about reforms that increased efficiency, made their jobs more satisfying, and eliminated donkey work. But others were upset, Calhoun says, and for a while job turnover was too high.

Work simplification is flowing through other departments at

Intel. Jay R. Elliot, who moved over recently from IBM to become personnel manager for California employees, jumped at the chance to streamline his operation. He says, with typical Intel jauntiness, "I want to run the best personnel department in the world." Workers in Silicon Valley are highly mobile, and piracy among competitors abounds. Elliot's employment office has to find replacements for a third or more of Intel's 6000 California employees every year. Nevertheless, it has cut the personnel staff from thirty-two to twenty people, thanks to simplification of the hiring process.

The personnel records department, described in 1979, by supervisor Karen M. Fox as "most disorganized, unbelievably unproductive," has been able to cut its staff from fourteen to seven. A whole series of gruesome chores, such as alphabetizing five hundred "personnel action notices" a week by hand, has been eliminated or computerized. The department finished 92 percent of its work on time in April 1981, compared with only 72 percent in February. All told, the personnel office in Santa Clara has been cut from one hundred twenty to eighty-eight people.

The supply offices, which provide pencils, paper, and typewriter ribbons, have achieved the most sweeping simplifications of all. Back in its early days, Intel got into the bad habit of handling all supplies as if they were the gold used in making microprocessors. "We dealt with small, precious products and treated everything as if it were small and precious," says Thomas L. Hogue, director of materials. That's why it took twelve pieces of paper and ninety-five steps to get a pencil out of a warehouse.

After hacking off some of these steps through normal job simplification, Intel decided on a drastic amputation by switching to "stockless inventory." Instead of keeping a warehouse full of stationery supplies for the Santa Clara offices, it turned the job over to a stationery company. Now when someone needs pencils or typewriter ribbons, he just fills out a requisition and drops it in a basket in the reception area. The stationer picks up the slips and delivers the item within twenty-four hours to the person who ordered it. The slightly higher cost of stationery is more than offset by reductions in supply workers and warehouse space.

NO BETTER IDEAS

Job simplification seems to be working. "It's the easiest way to save money," Hootnick says. Andy Grove, however, offers a caution: "This is a very, very new discipline, and we have no idea in detail how it's going to work." But then he adds, "It's going to play a big role for us. We don't have any better ideas."

Meanwhile, the pressures to do better are increasing. The growth of demand for Intel's products slackened in 1980, and cheap chips are pouring out of competitors' factories. The company's pretax operating margin took a disastrous dive to 1 percent early in 1981. In one year the price of a sixteen-bit programmable memory chip dropped from $25 to less than $7. That kind of market action puts a premium on administrative efficiency.

Nevin, who has been promoted to director of computer operations but retains overall responsibility for productivity, faces the challenge of involving departments such as sales and engineering, which haven't yet been seduced by simplification. That can't be rushed. In a company run by participative management, it wouldn't do to force simplification down anyone's throat. "So far," Hootnick figures, "we have only hit fifteen to twenty percent of the administrative staff with really active programs, and we have seen only about three to four percent of the total effect we could get." Then, paraphrasing his personnel chief, he adds, "Our goal is to make this the best-administered company in the world by 1984—and to be able to prove it."

June 29, 1981
Research associate: Anna Cifelli

8

CAN DETROIT CATCH UP?

CHARLES G. BURCK

The U.S. automobile industry has come through a difficult decade, with challenges bumper to bumper, but its biggest battle is just beginning. Since 1980 the industry and its analysts have decoded, with mounting anxiety, the most dreadful secret of all about the Japanese competition: they have an enormous advantage in their lower manufacturing costs. Estimates of the differential range from about $1300 to $1700 for a comparable subcompact car delivered at a U.S. port. Since shipping and duty charges come to between $400 and $500, the Japanese appear to be building their car for something like $2000 less than U.S. manufacturers.

Unless the U.S. auto industry can reduce this cost disadvantage, its future is in peril. Detroit has built up enough small-car capacity to meet just about any likely demand once the market comes back, and is quickly catching up with the Japanese in product quality. But neither of these accomplishments will help much if the Japanese can easily undersell American cars.

It seems incredible that automakers are only now starting to take the full measure of their Japanese competitors. "We should have been systematically monitoring that competition for the past ten years," says Robert E. Cole, director of the Center for Japanese Studies at the University of Michigan and an authority on differences between the two auto industries.

Detroit got its first inkling of the Japanese cost advantage nearly five years ago but didn't find much cause for alarm. The differential represented only a form of price discipline at the low end of the market, and the industry was happy to concede much of that to imports anyway, at least temporarily. As recently as 1978, Ford's main problem was persuading customers to order six-cylinder engines instead of V-8s so it could meet federal fuel-economy regulations. The center of the market seemed secure, insulated by the American preference for larger cars.

But the 1979 oil shock drastically shifted the market toward smaller cars. The Japanese, moreover, were gearing up to broaden their own product mix. Today Japanese automobiles compete in 60 percent of the U.S. market. Their cost advantage suggests that they can price their products pretty much as they wish.

Labor costs are an important part of the Japanese advantage: auto-industry wages in Japan currently run about eight dollars an hour below those in the U.S. (see chart). Belatedly, U.S. labor and management are moving to undo the damage they have done themselves through such indulgences as automatic wage gains not linked to productivity. The 1979 contract, which continued down that road despite Detroit's worsening competitive situation, stands as one of the industry's last great monuments to complacency.

Besides trying to tighten up on wage costs, automakers are reducing the substantial managerial fat they have accumulated over the years. GM, which in earlier days virtually guaranteed its loyal managers lifetime employment, has embarked on a remorseless pruning. During the past two years, Ford has cut four billion dollars out of its overhead, partly by reducing the salaried work force in its U.S. automobile business by some 25 percent. Chrysler has cut its white-collar staff just about in half.

Together, though, the pay differential and excess management overhead probably account for not much more than half of Detroit's cost handicap. Like U.S. manufacturers in general, the auto industry has fallen far behind the Japanese in productivity. The purest measure appears, with depressing clarity, in the difference in worker-hours per car. The most exhaustive and detailed analysis

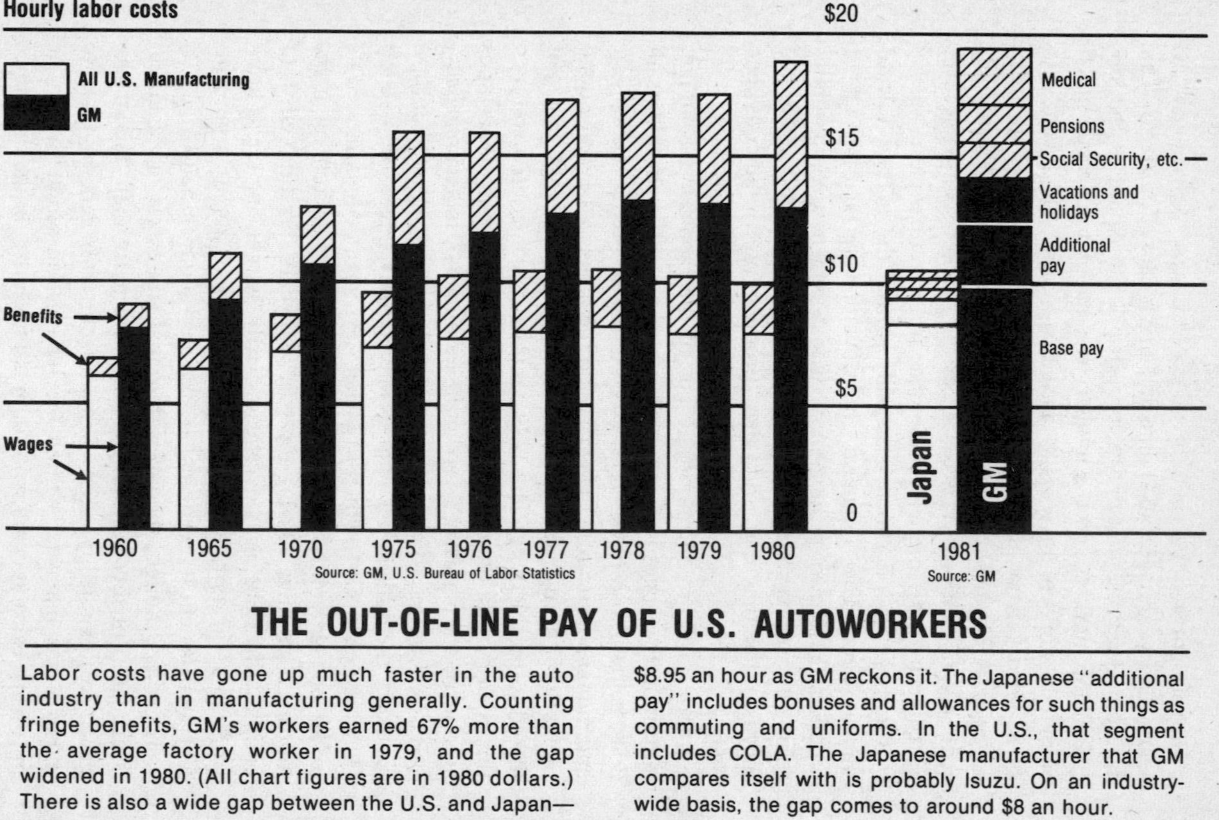

THE OUT-OF-LINE PAY OF U.S. AUTOWORKERS

Labor costs have gone up much faster in the auto industry than in manufacturing generally. Counting fringe benefits, GM's workers earned 67% more than the average factory worker in 1979, and the gap widened in 1980. (All chart figures are in 1980 dollars.) There is also a wide gap between the U.S. and Japan—$8.95 an hour as GM reckons it. The Japanese "additional pay" includes bonuses and allowances for such things as commuting and uniforms. In the U.S., that segment includes COLA. The Japanese manufacturer that GM compares itself with is probably Isuzu. On an industry-wide basis, the gap comes to around $8 an hour.

available, made by a former Chrysler manufacturing executive named James E. Harbour, suggests that the average Japanese compact or subcompact is assembled in fourteen worker-hours, against thirty-three for a comparable U.S. car. Building the Japanese engine requires 2.8 hours against 6.8 for the U.S. engine; stamping out body parts, 2.9 hours against 9.5. All told, the evidence indicates that the typical Japanese manufacturer builds cars with about half as many hours of labor as U.S. manufacturers.

Efficiency in the use of labor goes a long way toward accounting for Japan's low costs. The precise extent of the advantage cannot be calculated—reliable figures are not available. "There's a great deal we still don't know about costs," acknowledges Professor William J. Abernathy of Harvard Business School, who headed a National Academy of Engineering group that studied the two industries. Still, no one who has looked carefully at the Japanese automakers disputes the basic magnitude of their cost advantage.

THE MAN WHO WARNS DETROIT, "THIS IS WAR"

Andrew C. Brown

A reporter looking into the auto industry these days will almost certainly hear of James E. Harbour. He has a small consulting business—five people in all—but his name is likely to come up in almost any discussion of Detroit's inability to meet Japanese competition. Harbour, as one executive puts it, "brought the Japanese advantage into focus."

Until 1980 Harbour was little-known outside the Chrysler Corporation, where he spent twenty-three years, five of them as director of corporate manufacturing engineering. When the company down-sized its management, he left, started a consulting firm in a suburb of Detroit, and undertook to study auto-industry productivity. He already had an impressionistic grasp of Japanese-style manufacturing from his contact with Mitsubishi Motors, in which Chrysler has a 15 percent interest. "I watched that animal become so productive, it's disgusting," he says, "but I never put it all together. You can't analyze Japan from a macro standpoint; you have to do it plant by plant."

Analyzing data from automakers and government agencies, Harbour came up with estimates of the number of hours of labor that go into American and Japanese manufacturing processes. His figures—bad news for Detroit—made a strong impression on people in industry. So did his estimate that Japanese automakers have a $1700-per-car cost advantage over their American competitors. Harbour attributes most of the gap to Detroit's adversarial labor relations, excessive inventories, lagging productivity, and inferior quality performance. In short, poor management.

The study of Detroit's competitive weakness was hardly a dispassionate exercise. "I was twenty-three years in this industry, ten years on the bonus roll," says Harbour, 54. "The toughest thing to do is admit you're part of the problem." While auto executives quarrel with some of his numbers, they generally agree that Harbour is on the right track.

"Who gives a damn what the right number is?" Harbour snorts. "It's a vast difference, so let's go to work on it." He envisions a taut, harmonious industry running on a Japanese-style "just-in-time" production system. "To learn how not maintaining inventory helps to improve quality is so enlightening," he says. "Think how efficient our system would be if we all lived hand to mouth."

Harbour's clients are mostly small auto-parts suppliers—companies that don't maintain in-house staffs to study productivity—but he's going after big clients in a linkup with Arthur D. Little, Inc. "I'm only worried there's not enough time," he says. "I firmly believe that the Japanese will stop at nothing in this market. I just can't tolerate the fact that, while we're down in the dumps, they're kicking us in the teeth. This is war."

This formidable gap is not the product of leading-edge technologies—so far, at least, Japanese automakers are generally no more automated than those in the U.S., and in some cases, less so. "The latest engine plant in Detroit, with computer controls and laser measurements, is more sophisticated than anything we've seen in Japan," says Abernathy. "But a Japanese plant with fifteen-year-old equipment has half the labor."

The Japanese work smarter. Manufacturing techniques and methods, ranging from materials-handling systems to maintenance procedures, all reflect the careful attention of both managers and workers to the details of production. Indeed, decisions in Japanese companies about such things as production planning, quality control, and industrial relations are considered worth top management's attention as much as decisions about capacity or technology.

One thing the Japanese do distinctly better than their American competitors is handle inventories. U.S. companies carry large amounts of inventory to buffer the unevenness of production scheduling. The Japanese, in contrast, keep production inventories lean by means of careful scheduling and planning.

The system is called, appropriately, "just-in-time." Production schedules are fixed weeks in advance. Materials and parts arrive at work stations only as they are needed. Seats, for example, are delivered in a supplier's truck every few hours, and workers pull them out and feed them directly onto a conveyor. Because the supplier has loaded them according to the factory's schedule, the seats arrive at the installation point in coordination with all the other parts of the cars for which they are intended: a series of black seats for cars with black interiors, followed by a brown series, and so forth. The system does not provide much flexibility in production schedules, and Japanese automakers do not offer as many combinations of options as U.S companies, but the cost savings are substantial.

The most obvious benefits of just-in-time are the direct savings from having less inventory on hand. By Harbour's estimate, U.S. automakers carry $775 in work-in-process inventory for each car they build, while the Japanese carry only $150. Their interest expenses are lower and they need less warehouse space—and less plant space too. An assembly plant with a capacity of one thousand cars a day, says Harbour, will typically measure 1½ million square feet in Japan, as against more than two million for one in the U.S.; it is correspondingly cheaper to build, heat, and maintain.

But just-in-time has a larger purpose: to expose the manufacturing process to clearer view. When most in-process inventory is removed from a plant, supervisors can see workers and workers each other. Production deficiencies stand revealed—the machine that breaks down every three hours is not masked by a buffer pile of inventory; the operator who lets defective parts slip into the system has nowhere to hide his mistakes. The result is more up-time from machinery, less scrap, and more efficient use of workers' time.

Just-in-time exemplifies a production system geared to preventing defects rather than detecting them. The distinction is fundamental to Japanese factory operation. The Japanese regard waste, whether of material or human resources, as an abomination. Emphasis on product quality reduces waste: less scrappage, fewer rejects, fewer returns. And, happily, the Japanese have found that high quality and low costs are not incompatible goals that can be resolved only through trade-offs—generally the U.S. view—but are linked and complementary objectives.

To transplant Japanese manufacturing techniques such as just-in-time, Detroit has to understand the underlying principles of the system. Above all, it has to appreciate how much those techniques rely on a work force that enjoys a sense of participation with management. The Japanese, for example, work constantly to improve productivity through attention to details. Robert Cole observes that U.S. managers tend toward a "big-bang theory" of raising productivity with major technological investments and management reorganizations. The Japanese have discovered that more gains can be had from continuous, small-scale improvements that require the active participation of the work force in the enhancement of productivity.

Japan's now-familiar quality circles elicit ideas and suggestions from workers, but what's more important is that, in general, workers get a high degree of responsibility and respect on the job. A typical Japanese assembly plant gets by with fewer than half the inspectors needed in a U.S. plant. For one thing, the system produces fewer defects. Each worker, moreover, is himself an inspector and will not pass along a defective part.

"SEVEN UNIONS, ALL RADICAL"

The notion lingers in the U.S. that the Japanese system operates according to inscrutable and inimitable cultural characteristics. But Japanese successes in manufacturing operations overseas, with non-Japanese workers, should dispel any such idea. Two years ago Chrysler sold its Australian subsidiary to Mitsubishi. Before the sale, says R. W. Rawlings, Chrysler's director of productivity planning, the business was "in terrible shape—Australian trade barriers had created probably the most inefficient auto industry in the world." Chrysler Australia employed some 6800 people to turn out a mere 228 cars a day. Labor relations were dismal then ("typically British," says Rawlings, "with seven unions, all radical"). But when he visited recently, Rawlings found the company turning out 214 cars a day with fewer than 3500 people. Mitsubishi did it primarily by setting up a just-in-time system—inventory levels today, says Rawlings, are down to those typical of a Japanese plant. In addition, it totally reshaped labor relations and virtually guaranteed its workers permanent employment. "When we walked through the plant and stopped to talk to hourly workers, we found them enthused about the management, the company, the car," marvels Rawlings. "An amazing thing to see."

Nissan, starting with a green field rather than an existing plant, expects that the pickup trucks it will begin building in Tennessee next year will be competitive with those brought in from Japan. Nissan generally relies more on automation than Toyota, and the Tennessee plant will have 220 robots, even more than the company's highly automated truck plant on the island of Kyushu. But much attention will be paid to management procedures and practices. "Workers and managers will participate more with each other than in American companies," says Marvin T. Runyon, the U.S. subsidiary's president and chief executive. "The job descriptions will be fairly all-encompassing: hourly workers will do the job, the inspection, and the routine maintenance of the equipment. And the maintenance technicians will be qualified to do everything to fix a problem—electrical, mechanical, and so on." Shuichi Yoshida, Nissan's U.S. vice-president for quality assurance, pays Runyon, a former Ford executive, a compliment that speaks volumes about the qualities Japanese automakers prize in a manager. "Mr. Runyon is a very suitable person for this," says Yoshida. "He's not one of the financial people. I trust him—he was once a worker, you know."

Over the years Detroit has been maddeningly resistant to learning from outsiders, but that attitude is clearly changing. As Robert Cole puts it, "The recognition is sinking in fairly rapidly that business as usual won't get us through this crisis." Knots of U.S. auto executives and union officials now stroll, gaping, through Japanese plants, poignantly reminiscent of the Japanese who swarmed on the Detroit grand tour two decades ago. Most of the Americans are returning as converts. Anyone talking with a good production man who has visited Japan is likely to be struck by his enthusiasm for what he has seen. Most behave like men who have been waiting for just such a movement to come along. "I've never seen a more exciting time in the auto business," exults George Butts, Chrysler's vice-president for quality and productivity.

TOWARD AN AMERICAN VERSION

Even the enthusiasts recognize that outright imitation is not the way to go. "When we came back from the last trip to Japan," Butts says, "we asked ourselves, 'What can we move on quickly to implement in an American way?' We can't do everything the way they have done it. We have to do it our way—and we have to do it faster."

Detroit has already taken quite a few steps toward creating an American version of the Japanese system—including programs that go back several years. Some managers, moreover, are moving pretty fast. William J. Harahan, Ford's director of manufacturing planning, points to the company's progress in adopting a version of just-in-time: "We've dramatically reduced our inventories worldwide, and we've saved hundreds of millions of dollars. In the past it wasn't unusual to have forty days' supply of in-process inventory at an engine or transmission plant. We've cut that in some cases to two or three days."

For the past year, assembly schedules at the Escort/Lynx plant in Wayne, Michigan, have been set as far in advance as possible—the goal is four weeks. Each night Wayne calls the Dearborn engine plant, twenty miles away, with the next day's assembly lineup; Dearborn then ships the engines out in the right order. In the interest of simplifying assembly, Ford has also been combing through its options twice each model year and getting rid of those ordered by less than 2 percent of the buyers—special axle ratios or seat fabrics, for example.

GM is doing the same sorts of things, but it is also putting heavy reliance on automation for future gains in efficiency. Says Alex C. Mair, vice-president for technical staffs: "With the technologies we've developed in our manufacturing and facilities research groups, we could completely automate car and truck manufacturing for all practical purposes." It is not possible to go that far fast, but GM is moving in that direction. The company plans to increase its robot corps from 1400 to 14,000 during the decade. "By 1990 we'll probably have a work force that is half the size of the one we had in 1980," says Mair.

All the automakers are revising their relationships with outside suppliers. In the past Detroit shaved costs by ruthlessly pushing its suppliers to sacrifice their own margins. These policies helped the automakers' short-term earnings but hurt over the long run, by trading away quality and reliability. The Japanese, in contrast, are usually linked to their suppliers with long-term contracts (and often equity stakes as well) and work closely with them on product design and quality control. The close ties also build the cooperative relationships that make just-in-time production possible.

U.S. automakers are now signing contracts with suppliers for up to five years, especially for high-technology products that may require the supplier to contribute capital and research and development expertise. Suppliers benefit not only from the stability but also from what they learn by working closely with the factories. GM's Guide division supplies some Japanese automakers with headlights for U.S.-bound cars, and the Japanese have made a number of suggestions for quality-control procedures that Guide has accepted eagerly. "Other customers," says Boyd L. Bonner, Guide's general sales manager, "are interested mainly in price, ability to deliver, and whether the product meets government and industry standards. Nissan wants their unit to be better than the same unit we sell to Toyota. Toyota wants it to be better than Nissan's. Both want it to be better than the one we sell to Chevrolet."

Even where they do not have long-term relationships, automakers are becoming more active overseers of their suppliers. Ford has begun sending its own quality-assurance people to visit supplier plants on a regular basis. Where quality falls short of standards, the Ford people offer advice and help. The supplier who can't or won't take it is dropped.

Following another Japanese pattern, U.S. automakers are paying more attention to designing new cars according to the twin objectives of high quality and low cost. "The two do go together," says GM's Alex Mair. "So far I've seen no instance where it isn't true, which is an interesting new finding." But to get the two going together, the industry has had to reorganize its technical staffs. In the past, designers generally developed a new car design and consulted with manufacturing engineers only after the car had been approved by top management. Today manufacturing engineers work with the designers—along with engineers concerned with service and maintenance—from the beginning.

Perhaps the most promising development of all is that these programs are accompanied by fundamental management changes. GM's staff cuts reflect not simply an economy move but an effort to streamline the corporate hierarchy and make management more effective. One important aspect of the change is that the divisions are being prodded to reshape their own managements. Over the past two years the company has served notice on all its own components divisions that they cannot consider their roles secure just because they're part of GM; they will have to compete in price, quality, and service with any potential outside supplier. "One of our business goals," says vice-chairman Howard H. Kehrl, "is to do a goddamned good make-or-buy decision on everything. We've got some stuff in-house we shouldn't be doing—or shouldn't be doing like we're doing it."

One of the most important elements of GM's efforts to improve productivity was actually started a decade ago—the quality-of-worklife concept, which lays great stress on making the worker a full participant in improving production techniques and practices. Quality of worklife is firmly established in more than half of GM's plants. Recently Alfred S. Warren Jr., vice-president for industrial relations, organized a group to think strategically about industrial relations, with members drawn from staffs as disparate as personnel and financial. The group is now preoccupied with the UAW contract, but Warren expects that its long-run purpose "will be to ask questions like, What should industrial relations be? and, What is the role of the hourly employee?"

Ford, with its highly centralized management style, has been able to move faster than GM in implementing some of the same ideas—its so-called "employee-involvement program," for example, is under way at two-thirds of its plants. At the same time, there are some signs that chairman Philip Caldwell is trying to break the rigidly hierarchical, risk-averse style that has burdened the company for most of its history. "We're increasing delegation down the line and letting people write their own records," says Caldwell. "In the supervisory ranks, why should one person supervise two people? Why not five or six?"

A recovery of the U.S. auto market is likely to restore an appearance of health to the industry. Automakers are poised to profit from even a modest sales upturn: by the estimate of security analyst David Eisenberg of Sanford C. Bernstein & Co., the industry has reduced its break-even point from 11.9 million to 10.1 million vehicles over the past three years. Moreover, volume increases should bring sharply higher productivity, helping somewhat to narrow the Japanese cost advantage.

The survival of the bigger-than-small car is a plus for Detroit.

Sales patterns of recent months have shown that Americans still prefer larger cars than do Europeans and Asians. The Japanese are growing stronger in markets for larger and more profitable cars, but the U.S. industry has a great deal more capacity available to it—much of it in old, largely depreciated rear-wheel-drive technology. Modestly lightened, reskinned, and equipped with fuel-efficient engines, some of those vehicles could go on making money for Detroit for some time to come.

To thrive or even survive over the longer run, Detroit will have to narrow the Japanese cost advantage. James Harbour, for one, thinks that over the next five years U.S. automakers could eliminate around two-thirds of their present manufacturing cost disadvantage—apart from whatever savings can be squeezed out of wage costs. That assumes auto company managements will carry through the changes they have set in motion. Chrysler's Australian experience does show that a management that knows what it's doing can make remarkable improvements in only a couple of years. In its competition with Japan, the U.S. auto industry has what has been called the advantage of backwardness. Potentially it can achieve large gains in efficiency for a while because it has has been doing so many things inadequately—just as a developing country with a competent government and sensible economic policies can grow much faster than the U.S. or the built-up economies of Western Europe.

The Japanese, of course, will not be standing still, but their own productivity gains will come much harder during the 1980s, mainly through costly automation. And volume growth, undoubtedly a major source of productivity gains in the past, will slow down. The Japanese auto industry's volume has been expanding at a compound annual rate of almost 11 percent for the last five years, but David Eisenberg estimates it will grow at 7 percent or less between now and 1985.

So Detroit may have a future after all. The U.S. automakers have just about exhausted their margins for error and bad luck, but they have a clear view of their competition's strength and of how, in general terms, they can respond. If the managers can work smarter than they did in the 1970s—a lot smarter—they have an opportunity to pull off one of the great business turnarounds of the century.

February 8, 1982
Research associate: Andrew C. Brown

9

WHEN WORKERS MANAGE THEMSELVES

CHARLES G. BURCK

The great industrial parable of the 1980s is the factory-salvation tale. An increasingly common story, it generally describes the miraculous turnaround of a faltering operation though some form of new employee-participation effort. Those who narrate the tale, whether members of management or labor, often do so with the fervor of religious converts. Hard-boiled foremen tell how they learned, to their delight and astonishment, that the workers had countless ideas for improving the operation; once-militant shop stewards stand up at plant meetings to affirm labor's commitment to high productivity and profits. The results are plain to see: productivity is up 20 percent, grievances are down 90 percent, and the plant is making money for the first time since the bottom fell out of the widget market.

Yet tales of salvation, however edifying, cannot fail to raise suspicions. The enthusiasm of the born-again is apt to reflect the ecstasy of the epiphany as much as the facts surrounding the event. The corporate champion of a quality-of-worklife program, as schemes for broadening employee responsibility are often called, is counting on it to advance his career. It damn well better be called a success!

The growing popularity of employee-participation programs

isn't really in doubt. After a decade of experimenting and testing, the basic principles have passed from theory into practice in hundreds of factories and offices across the land. But equally clear is that mystery, debate, and confusion remain over whether these ideas are really paying off when measured unemotionally by the calipers of corporate benefit.

One difficulty is that even seemingly precise yardsticks for measuring accomplishments, such as higher output per employee, are not wholly reliable. Few companies can account fully for all the variables that may have affected the outcome of a change in the way people work together. The first year or so of good results might reflect nothing more than the tendency to work harder under a new spotlight. Gains may also be attributable to progress along a learning curve. Or to the threat of a plant closing, which can produce productivity miracles.

Since the early 1970s, scores of companies have flirted with innovative work arrangements on the plant floor or in the back office only to come up empty-handed. If all had gone beyond the experimental stage, employee participation would probably be the norm in the U.S. today. Instead, in many instances programs never spread beyond the plant or division in which they started, and most of these "encapsulated" programs, as social scientists in the field call them, were eventually scaled back or abandoned.

Many failed because they *were* experiments, conducted in the infancy of the social-science disciplines of the workplace. More often, the champions of employee participation underestimated organizational resistance to the idea. Lockheed was one of the first companies in the U.S. to use quality circles, but middle management was not persuaded that they made sense. When the idea's partisans left the company in 1976 to become consultants, the program languished; Lockheed is now attempting a revival. General Foods got attention in the early 1970s when it opened a pet-food plant in Topeka with a highly participative system. But the operation offended traditionalists at General Foods and, by most accounts, remains encapsulated. (The company won't discuss the matter.)

BOMBARDED FROM BELOW

The trouble with the plant-salvation parable is not that it promises too much but that it makes conversion look simple and quick, when it is neither. The record suggests that many companies charging off today to make themselves more participative will return tomorrow defeated and bewildered. If hourly workers take the idea of participation seriously, their foremen or supervisors are going to be bombarded with questions about whether established methods are the best way to get the job done. Pressures on the supervisors, in turn, will have repercussions in the layers above. Managers at all levels will have to discover new ways to assert leadership. Those who cannot take command of the forces crashing through their organizations are likely to find their organizations coming apart.

Nonetheless, the movement has drawn strength from the stature of notable corporations that have been developing shop-floor participation for years and have made the concept part of their management philosophy—companies as diverse as General Motors, Polaroid, Dana, Herman Miller, TRW, and Procter & Gamble. These companies generally did not set out with rulers to measure productivity gains—mainly to avoid giving workers the impression that productivity was all they cared about. But they saw progress in such things as reduced absenteeism and higher product quality. They also perceived benefits that cannot always be measured, such as the unknowable savings from an idea generated by a quality circle that prevented a product recall. And they have seen that employee participation has, over the long run, made entire organizations more effective.

Some of these companies started with the simplest forms of employee participation and found that even these represented a basic challenge to convention. Hundreds of corporations and government bodies in the U.S. are now using quality circles, and the number is growing almost daily. They originated in Japan as relatively narrow programs focused on statistical quality control, but in the U.S. they have evolved into broader participative mechanisms. They may deal not only with the technical details of quality control but also with environmental factors that affect the work—say, factory lighting or seating—and even with changes in the production process or in work schedules.

Quality circles at Northrop Corporation's Aircraft Division near Los Angeles have brought workers' ingenuity to bear on a host of quality and operating problems. Northrop is convinced that the programs have brought huge payoffs. The fifty-five or so blue- and white-collar circles now operating at the Aircraft Division are concentrated in the 747 operations. "During the two years that we've emphasized that it is *us* rather than *thee and me*, the cost of the 747 unit we're delivering to our customer went down fifty percent," says Bev T. Moser, vice-president for commercial operations at the division. Costs might have dropped significantly without quality circles, he allows. "But to put that fifty percent in perspective, we had to quadruple our work force during the period as business picked up, and had to train people who'd never seen an airplane."

Results like that will not be realized by credulous (or merely lazy) managers who tend to utter magic phrases over problems and declare them solved. A small survey conducted early in 1981 for the International Association of Quality Circles, a group of professional organizers and trainers, revealed that many circles are nothing more than monthly supervisors' meetings, or traditional project committees set up to deal with problems identified by management. Even a genuine effort to organize a quality-circle program is apt to founder without a great deal of preparation. Left to their own devices, members are likely to spend fruitless hours in meandering discussions; eventually, either management or the group itself grows discouraged with the pointlessness of it all, and the program is abandoned.

In a well-organized program like Northrop's, team members are given a good dose of training in the basic techniques of problem-solving. They learn how to gather and analyze data, weed out trivial issues to focus on major problems, generate innovative ideas in brainstorming sessions, forge consensus decisions, and communicate effectively. At least one team member—usually, though not always, the supervisor—gets extra training in leadership. And any well-run program will have one or more people trained as "facilitators," who help organize groups and get people who were accustomed to performing isolated jobs on an assembly line to begin thinking, talking, listening, and caring as members of a team.

Once in place a quality-circle program needs tending, like a favorite rosebush. As elementary as that might seem, it came as something of a surprise even to Hewlett-Packard, an electronics company celebrated for its creed of accessible management and respect for the worth of its workers. In July 1981, two and a half years after setting up its first quality circle, the company had five hundred of them. "But it's not a simple program," says Fred Riley, product-assurance manager for the manufacturing division. "If it were, it would have been done long ago. In many cases we have a tendency to start a program, walk away, and assume it is going to run by itself. You can't just do that. It is going to take a lot of work for a long time."

NEW TEETH FOR THE GEAR

A company takes a major leap when it goes beyond the quality circle to a broader participation philosophy, in which the employees plan and schedule the work and, in some cases, redesign their jobs. No ground less fertile for such seeding might be imagined than the Chevrolet Gear & Axle plant in Detroit, a sprawling seven-building complex that grew old with all the ugly aspects of traditional union-management divisiveness, poor communications, and excessive hierarchy. By the late 1970s the Gear, as workers call it, was selling several hundred million dollars' worth of components to various GM divisions, but it was in serious trouble. The planned switch to front-wheel drive for the X and J cars threatened one of its major product lines, conventional rear axles. The division lacked any strategic vision of its future.

Extraordinary as it may seem at as huge an enterprise as GM, Ray McGarry, the plant manager, took it upon himself to set about totally reorganizing the management structure of the division. From Harvard, he brought in Paul Lawrence, a noted specialist in organizational planning, who proposed major changes aimed at compressing the hierarchy and pushing responsibility for decisions down as far as possible in the organization. At about the same time, the Gear's top managers attended a GM corporate quality-of-worklife presentation and came away with a new conviction. As Bob Vervinck, the plant's personnel director, recalls: "We realized that you could change management, but you'd never have full success without the full participation of the union and the employees."

The United Auto Workers has been the leader among labor unions in embracing employee participation. In 1979 representatives from management and Local 235 retreated to a nearby conference center for two and a half days of meetings to discuss the possibilities. So important did the issues seem to both sides that contract negotiations then under way were suspended and adversarial roles laid aside for the duration.

Despite predictable differences of opinion, a surprising consensus emerged. Both sides wrote down their objectives for the future of the Gear. When the lists were posted side by side, says Vervinck, "An outsider would have been hard-pressed to pick who developed which." The shared goals ranged from making the jobs more satisfying to producing quality products and a profit. Coming out of the meeting, both sides agreed to move forward with a program guided by a committee made up of four union representatives picked by management and four management members chosen by the union.

The Gear used to be a classic assembly-line operation, with each worker at his post on the line performing an unvarying task, doing what he had to and avoiding what he could. Today it has some sixty-five teams of hourly workers, with a total membership of between 600 and 650 people (about 12 percent of the work force at the Gear). While in the past a worker might have stood in one place all day tightening bolts on a rear brake subassembly, he is now responsible with other team members for the production and quality of an entire brake system. All the workers are capable of doing one another's jobs, and, as Doug Latkowski, a quality-assurance supervisor, observes, "All of us are inspectors." Team members, obsessed with quality, have pleaded with plant engineers and harangued outside equipment suppliers over matters they thought might compromise their standards. Chuck Twymon, who sets up and adjusts machines on the line, cheerfully asserts: "I'm not going to see my job go down the tubes just because some engineer is too bull-headed to listen to me."

Managers who want their workers to act like this must make no mistake about the months needed for planning and the hundreds if not thousands of man-hours required for training. When GM expanded a quality-of-worklife program at its Tarrytown, New York, assembly plant in 1977, every manager there spent three days learning about the principles and goals. Some three thousand hourly workers received twenty-seven hours each of technical and problem-solving training—at a cost of more than $1 million (including wages to the workers and their replacements while they trained). In starting up a truck-assembly operation in Dayton, Ohio, in 1981, the company gave each worker full pay for forty hours of training—half of it devoted to the team-building and problem-solving skills needed for participative management.

Bringing financial information down to the shop floor is a major step in bridging the gap between management and labor; more than any other single act, it makes the goals explicit and the nature of the partnership concrete. At the Gear, managers tell workers about the plant's direct labor costs, scrap costs, and profit (or loss)—and how these measure up against goals. Not even the foremen would have been privy to such information at GM a few years ago. The benefits, to GM's way of thinking, outweigh any harm that might come from revealing competitive information.

For many managers, this kind of disclosure still seems startling, if not subversive, but it has worked for years in companies such as Dana Corporation, a $2.5-billion-a-year maker of industrial equipment and motor-vehicle parts, and Herman Miller, Inc., the systems-furniture maker. Both use a well-established participation mechanism known as the Scanlon Plan. Created in the 1930s by a labor leader and cost accountant named Joe Scanlon, the plan shares productivity gains with the workers Labor and management agree on a figure for "normal" labor cost—those in the period preceding the plan. Each month that costs fall below that norm, the difference is assigned to a bonus pool that is split among the company, the plant's workers, and its managers. In particularly successful cases, employees can earn 10 to 25 percent more a year in bonuses.

Dana's experience has shown that no single formula works the same way in all applications. In different plants the Scanlon Plan is implemented in different forms, ranging from quality circles to fully cross-trained work teams. Most remarkably. Dana has found that the financial motivation of productivity-gain sharing, the most attention-getting aspect of the plan, does not have the same importance to all workers. In some Dana plants, no bonuses have been paid for a year because of depressed market conditions. Yet the employees have voted to renew the plan, attend the monthly meetings, ask questions, and generate ideas for improving operations because they regard participation as the major benefit.

In letting plants develop their own variants of the Scanlon Plan, Dana avoided one of the great pitfalls of participative management. Academics, consultants, and experienced corporate line managers all stress that every program has to be tailored to the specific character of the operation that adopts it and the people who work there. All too often executives try to impose a single concept across the board. Thomas G. Cummings, a pioneering consultant and a professor at the University of Southern California's Graduate School of Business Administration, observes tartly: "There's no one I know who'll drive down the Harbor Freeway blindfolded, yet they'll implement these approaches without honestly assessing themselves so they can better match the system to the situation."

FACTORY-FLOOR VIRTUOSOS

The most far-reaching participation programs are apt to spring up in "green field" plants that are new from the ground up. TRW's

Lawrence Cable Division in Lawrence, Kansas, which makes cables for submersible oil-well pumps, has been run by work teams since the day it opened in 1976. "In the beginning we considered it an experiment," says the plant's ebullient and forthright general manager, Gino T. Strippoli. "But somewhere along the way we said, 'This is no longer an experiment; this is how we operate.' "

Lawrence's 130 nonunion employees are grouped in ten teams, ranging from five to twenty-eight members. Most members have no specified jobs; they can handle most of the functions required of their team. In many cases they can also take on indirectly related tasks, such as driving a forklift or handling materials. Says Strippoli, "As a manager, I love that. If you look at forklifts in traditional plants, they never have more than twenty or twenty-five hours of running time on the clocks at the end of the week. So what are the operators doing with those other fifteen hours? Here we don't have that. The team members are doing whatever's needed—running machines, removing reels, picking up scrap."

At Lawrence management assigns output goals while the teams set their own schedules and deal with matters ranging from quality problems to overtime. The teams are also responsible for training new members, appraising their performance, and recommending who should get raises.

The system is continuously modified. Workers have merged overlapping teams to increase flexibility, and at the moment are trying to cut down a team that has grown large and cumbersome. Originally raises were given only when workers mastered new pieces of equipment. The problem with that, says Strippoli, was "all everyone wanted to do was train for new jobs." Now raises are also awarded for increased proficiency in existing skills.

Strippoli acknowledges that managing such an operation is harder than managing a conventional hierarchy. Now, when he makes a decision, workers will call him on it if they think he's wrong. "I have to be right a hell of lot more often than I used to be." At the same time his job has been refashioned. Workers who have broad responsibilities for the product don't just stand around when the line goes down; when something goes wrong they fix it, leaving Strippoli time to think beyond minute-by-minute operations to matters that will improve performance in the future. "I really feel for the first time that I'm managing rather than putting out fires. The teams are putting out the fires way down in the organization."

Generally the managerial art is one of persuasion and direction. Norman St. Laurent, the manager of manufacturing engineering at Lawrence, recalls a problem the plant was having with a particular machine: "Since everybody is trained to have their own ideas, I had my idea, the first shift had their ideas, the second shift had theirs, and so on." The solution emerged only after St. Laurent called an inter-shift meeting and got the group to agree on a systematic trial of the ideas.

But persuasion has to be backed up with tough-mindedness, and participation can't be confused with permissiveness. Cummings of U.S.C. recalls a company he dealt with that failed in its efforts because the managers didn't assert authority. "They had misinterpreted participation and were getting anarchy instead," he says. Strippoli has no such problem: "The workers know that if I feel there's no payback to the company in the solution they arrive at, there will be a definite no. I'm not here to give away the store or run a country club."

Possibly the thorniest problem of all—and the reason a great many participative-management efforts fail—lies in middle management. Until recently, most architects of participation overlooked the middle manager and his problems. As Strippoli puts it: "Organizations assume that the manager is by nature intelligent, so what's the big deal about allowing him to see the light? The action is in getting *them*—the workers—to join *us*." Strippoli speaks from experience: "We went through a lot of grief because we didn't do a good job of training our management." Teams fired up with the participative spirit would bump into managers who thought of themselves as traditional bosses; both sides would come away bruised.

THE MESS IN THE MIDDLE

Any collaborative system brings substantial changes to the roles of the managers involved. If a group makes a decision that a foreman thinks is wrong, he cannot simply reverse it without destroying the participative compact. Yet he dare not let it stand either. Nothing in his experience has prepared him to sit down with the workers and talk things out.

The problems are no less severe at levels between the foreman and the plant manager. "In a traditional operation, I might give one of my managers a goal to reduce maintenance costs by thirteen percent," says Strippoli. "But I don't ask him what he's going to do to achieve that. I don't know who he's made angry, who he's lost, or what other problems he's created. Here, I still want him to cut costs by thirteen percent, but I also want him to treat his teams equitably so they keep operating efficiently."

For the manager caught in the middle, the answer is retraining in the arts of participative management. Companies like GM, which have firm plans to spread the participative philosophy eventually to all their operations, are just beginning to take on this task. They are also starting to assign weight to these skills when they judge candidates for promotion. But in many cases, they concede, the only solution may be to get rid of the people who cannot change. This is not a solution corporate leaders feel very comfortable about—especially those whose organizations trained their managers to employ authoritarian practices in the first place.

Companies that have had time to weigh the consequences of participative management are finding that it informs the entire corporate culture. When the system is used to the full advantage of both management and labor, blue-collar workers are no longer just workers: they become the lowest level of management. Because the company needs fewer administrators to supervise them, it can fashion a leaner and more responsive organization, with clearer and faster communication up and down the chain. Those who have glimpsed the full possibilities—and their number is growing—see employee participation as a form of managerial risk capital whose long-term payoff is a more effective organization.

July 27, 1981
Research associate: Faye Rice

10

WHAT'S IN IT FOR THE UNIONS

CHARLES G. BURCK

American labor leaders have by and large had little use for well-intentioned schemes to make them partners with management. In the words of Thomas R. Donahue, secretary-treasurer of the AFL-CIO, most such plans have been "the worst kind of nonsense," perpetrated either by "pied pipers who promised 'no supervisors, no assembly lines,' or romantic academics espousing European-style codetermination." Either way, says Donahue, "they set everyone's teeth on edge."

But a remarkable turnabout is under way. Quality circles, quality-of-worklife systems, and other techniques for enlarging employees' responsibilities are gaining favor among unions almost as quickly as among companies. Union leaders who used to worry that their members would accuse them of selling out to management have been surprised to find that this new species of cooperation has raised their standing with the rank and file.

The United Auto Workers, the United Steelworkers, the Communications Workers of America, the International Brotherhood of Electrical Workers, and the Telecommunications International Union—together representing a fifth of the nation's organized workers—have signed national labor agreements committing themselves to plans for bettering worklife. The United Rubber Workers, the Bakery, Confectionary, and Tobacco Workers International, the United Food and Commercial Workers, and others have quietly supported locals that try new ways of cooperating with management. Still others are watching cautiously, and nowhere are national union leaders speaking out against the concept. Even William Winpisinger, president of the International Association of Machinists, whose militant rhetoric about the folly of cooperating with management brings to mind the firebrands of the British labor movement, considers quality of worklife an issue apart—an improvement, he says, that he's been fighting for all along.

A good deal has changed since those days when the pied pipers came tootling through. Managers increasingly see that workers want to take their jobs seriously, and to be taken seriously by their supervisors. Today's quality-of-worklife concepts are far different from discredited schemes of the past; they offer clear benefits to both workers and management.

Adversity has helped make labor leaders readier to examine old orthodoxies. Unionized labor's share of the total work force has been declining for years, and its leaders are beginning to suspect that labor-management cooperation holds interesting possibilities for reversing the decline. Trapped, nonetheless, by years of adversarial posturing, many leaders remain uncomfortable about collaboration—particularly since experience has taught that management-inspired programs usually result in work speedups or manpower reductions.

During the early 1970s, for example, the major steel companies and the United Steelworkers agreed upon a bold plan to fight the growing threat of imported steel: joint labor-management committees would sit down together to reason out ways of raising productivity. Lloyd McBride, now the Steelworkers' president, remembers well the first of those meetings he attended, as a staff representative at the Granite City Steel Company.

"The management guys came in and said, 'Well, we want to talk about the productivity of this operation,' " McBride recalls. " 'Down in this department we could eliminate this job and that job, and over here we can have so-and-so double up because it's not very busy.' Our committee sat there for a minute, and then one of our guys got up and said, 'Well, that would create some problems for us. But the problems would not be so great if we could get rid of your brother-in-law down there who's not doing much, and Smith's cousin who's just sitting around and hasn't done a goddamned thing for years. If we could get rid of those problems, it's *really* going to improve our productivity.' " The nepotism McBride recalls in that plant wasn't a way of life in most steel mills, but the confrontational attitudes between labor and management were: the entire productivity program was soon abandoned.

At about the same time the United Auto Workers Union was trying to develop a plan jointly with General Motors to deal with roughly similar problems. But their approach was entirely different. GM had become increasingly concerned about the restiveness of its work force and aware that traditional ways of boosting productivity were of only limited use when workers came in late or failed to report at all. Its organizational-development staff, headed by a thoughtful psychologist named Delmar L. "Dutch" Landen, was determined to find ways to get workers more involved in their work. Irving Bluestone, then director of the UAW's GM department, saw an opportunity to advance his own long-held belief that unions could cooperate with management to advance the workers' responsibilities and stature.

Bluestone was a maverick within the union—his fellow officers thought he was making an enormous mistake—but he pressed on with his views. The company and the union began to talk about the possibilities in 1972, and as part of the 1973 contract negotiations they signed the first national quality-of-worklife agreement in the U.S.

That agreement has become the model for just about all the others signed since. Nowhere in it did the word productivity appear; the underlying premise was that management would seek its rewards from such improvements as higher product quality and lower absenteeism, which might be expected to spring naturally from a more satisfied work force. All the quality-of-worklife undertakings were to be strictly voluntary, and none would be used unilaterally to raise production rates or reduce manpower requirements.

The participative systems that evolved have been almost as diverse as the plants that practice them, but all have given workers more control over their jobs and opened up communications with supervisors. In most cases, workers have come to see themselves for the first time as participants in the enterprises in which they work.

The UAW's years of experience have done much to persuade others in the labor movement that joint programs with management can serve the interests of both sides. "They're a strong union and nobody's patsies," says John W. Shaughnessy, Jr., president of the Telecommunications International Union, whose 60,000-odd members are mainly Bell Telephone workers. "They've given quality of worklife a lot of credibility."

Like managers, labor leaders cannot always quantify the gains from such changes, but local disputes and grievances—a sure barometer of labor relations—are down dramatically where there are joint labor-management programs. Over the years just about every national settlement in the auto industry has produced protracted guerrilla warfare, as local units struggle—sometimes for weeks—to resolve disputes. But the 1979 settlement broke with tradition. At GM, a half dozen locals settled their problems even before the national contract was concluded—astounding, say bargainers on both sides of the table—and sixty-three others wrapped up matters simultaneously with the national agreement or a few days afterward. Virtually all were at plants with active quality-of-worklife efforts.

Labor leaders have been no less surprised than managers by the benefits of employee participation. Like the foreman who cannot believe that workers have useful ideas, the old-line militant shop steward cannot imagine how cooperating with management will advance his own career. But like the foreman, the union representative can become much more effective in his job once he learns how to take advantage of the new system. "Participation creates a new role for him," says Donald F. Ephlin, who is director of the UAW's Ford department. "The committeeman becomes more than a committeeman; he becomes a co-leader with management of a program that is bringing new status to the employees."

The typical shop-floor leader has to spend most of his time on grievances—answering members' calls, writing up complaints, and nagging foremen for weeks or months to get responses. In some plants grievance backlogs run into the thousands. "I tell people they should weigh them rather than count them," jokes UAW president Douglas Fraser. Most are generated by a misanthropic minority and are often inconsequential—a burned-out light bulb or an overflowing trash can—but they vent deeper discontents. Union representatives are often frustrated by such pointless exercises but are nonetheless compelled to fight for their constituents.

Where employees become participants, the grievance load invariably goes down. The general atmosphere of the plant improves, and many of the issues that might become grievances can be resolved informally. A broken toilet might not be repaired for a week when the union representative has to bring it to management's attention in writing. But if he can drop a word in the foreman's ear, and the foreman can dial plant maintenance on the spot, the plumbing gets fixed in a hurry.

The floor representative's own quality of worklife goes up under such circumstances—along with his rapport with the people he represents. "If I'm part of a process where we're talking things out, I'm spending more time getting acquainted with the members and building personal relationships," says William Horner, who retired from the UAW last year after serving most recently as Bluestone's administrative assistant for developing quality-of-worklife programs. Instead of writing up endless grievances or defending the minority of workers who account for the most absenteeism, the representative can find out more about the concerns of the majority and take them to management with more hope of a receptive hearing.

The practical results have done much to allay unionists' fears that employee participation would make the local leader superfluous. Quite the opposite: leaders in quality-of-worklife plants find themselves politically more popular than ever. To date, according to UAW leaders, virtually every slate of the union's officers who campaigned by supporting an established quality-of-worklife effort has won.

HONORING THE COMPACT

Many union leaders feared that labor-management cooperation would undermine the collective-bargaining system, the very foundation of the U.S. labor movement. After all, nothing could be further from the spirit of the bargaining table than a process in which both sides wrestle with mutual problems by putting their best ideas forward at the *start* of a discussion. Some unionists worried that management would use quality-of-worklife programs to chip away at benefits won through bargaining. Others took precisely the opposite view: they saw participation as a way to squeeze out gains they had failed to get during negotiations.

In practice, the lines between combat and cooperation have remained remarkably clear. Few managers or labor leaders in the auto industry have tried to abuse the compact, and none has succeeded. Yet neither side feels constrained to shelve issues that cannot be resolved cooperatively. GM and Ford are relentlessly pressing the UAW for wage concessions—and the UAW is mincing no words in telling the companies where to go. "When we get an issue where both sides feel they're right, we get our mad on," says Bill Horner. "At the extreme, we prepare for a strike or even go out. But someday it's resolved. Then the union and the company have to ask each other whether they want to continue the warfare or go back to cooperation. Invariably, we agree it's a better life to go back."

Old suspicions do not die easily, however, and unions that have recently committed themselves to pursuing shop-floor collaboration are proceeding with care. The agreement the United Steelworkers signed with the industry's nine major producers as part of the 1980 contract set up so-called labor-management participation teams that will function at the department level (rolling mill, shipping department, and so forth). These teams will have a charter to discuss practically any issue that might come up—with the important exclusion of grievances or matters negotiated in the contract.

Cooperation is an especially alien notion in the rough-and-tumble, highly authoritarian atmosphere of the steel mills. Here, only a few teams are functioning so far, and they are still regarded as experimental even though workers in nearby departments, who have seen the results, are clamoring to be included. As Lloyd McBride says, "It takes careful planning and training to prepare both sides for these new roles." Neither the industry nor the union wants to risk a reprise of the earlier fiasco.

Cooperative efforts in autos and steel were unquestionably spurred by the crises both industries have faced. No comparable threat from foreign competition has confronted the communications industry, but labor and management have found themselves in extraordinary agreement about their long-term needs. When the Communications Workers of America presented its quality-of-worklife proposal at the 1980 contract negotiations, says President Glenn E. Watts, "AT&T's counterproposal was so close to ours that we were quite surprised." Union members were increasingly frustrated by what they call job pressure. In its efforts to raise productivity, the company was scrutinizing performance through a microscope. Some techniques were indeed extreme—listening in on operators to judge their efficiency, or ringing absentee workers at home to confirm that they were sick when they claimed to be.

AT&T was no less eager to break with such procedures; it believed that making jobs more satisfying would yield productivity gains over the long run. "The company has made a decision to accept the union as a limited partner," says Michael Maccoby, director of the Project on Technology, Work, and Character in Washington, D.C., and a guru to both sides. "But in return it is getting a much higher level of cooperation, more flexibility as far as technological change is concerned, and the potential for huge savings by increasing managers' spans of control—because fewer of them will be needed to supervise the workers."

FIGURING OUT THE REALITIES

Thoughtful labor leaders are tantalized by the idea that participation may help them resolve some profoundly troubling union problems. No major labor leader in the U.S. today has illusions about the relationship between corporate profits and the well-being of his constituents. But down in the ranks, it's usually a different story: the myth remains widespread that corporate resources are somehow limitless, and labor's claim on them consequently open-ended. Labor leaders may have felt they could afford to encourage this nonsense for their own political ends during times of rising prosperity, but many are now faced with the unpalatable job of telling their members why demanding less will be good for them. They find it hard to explain the facts of life to the membership until—as with Chrysler—an afflicted enterprise is on the brink of collapse. Even then, leaders cannot do much to break the momentum of othodoxy; writers of the UAW newsletter, *Solidarity,* still churn out broadsides telling auto workers they are impoverished, despite what they hear from the "big-business-controlled media."

Many labor leaders now hope—though they do not like to say so explicitly—that members who become participants will figure out the economic realities for themselves. As McBride puts it, "Workers with more information will be able to make better decisions about union policies."

This indeed seems to happen where management has shared financial information with the workers, as at many auto plants, and where local leaders overcome their own fears about being accused of selling out to the company. "It's not getting in bed with management—that's the clarification you need to make," says William "Red" Hutchins, president of United Rubber Workers Local 87 at GM's Inland division in Dayton, Ohio. "We have to stop looking at whether we're afraid somebody will think we're taking a wrong position just because we want to see the company make money. If the company doesn't, it's hard for us to go back and negotiate increased benefits or wages." While Inland is not operating in the crisis atmosphere typical of many automotive operations these days, the union leaders are looking warily to the future. Dayton has lost its share of employers in recent years, and as Hutchins says: "We don't want to become another Frigidaire, NCR, or Dayton Press"—all of which have shut down plants in the city. "We're starting early to keep that from happening." The division's quality-of-worklife efforts began in 1980.

THE ART OF AMICABLE ORGANIZING

Opinions are still divided about whether employee participation is a threat or an aid to new organizing efforts. In some nonunion plants, management has used it to resolve discontents that might have brought in a union. Yet some leaders believe it can give unionism new leverage in attracting the unorganized. "Where you run into a 'no union' attitude, it's because the union is seen as a belligerent and aggressive outfit coming in to upset the established work relationship," says Shaughnessy of the Telecommunications International Union. The TIU's California local won thirteen elections in two years in Pacific Telephone Company offices by eschewing the classically confrontational rhetoric offered up by its rival, the CWA local. (Like AT&T operating companies, the union locals have considerable discretion about their own operating styles.) Pushing joint labor-management committees that will head off problems before they turn into grievances, the TIU organizers have won over workers who previously resisted representation.

For most union leaders, worker participation is still too unfamiliar to embrace. But many who want to learn more about it are seeking counsel from specialists in labor-management bridge-building, such as Michael Maccoby, Ted Mills of the American Center for the Quality of Work Life in Washington, D.C., and Jerome M. Rosow, president of the Work in America Institute in Scarsdale, New York. Rosow's institute in 1980 organized the Productivity Forum, which provides a neutral ground for corporate executives, labor leaders, and government and educational managers to meet and discuss measures for improving working relationships and productivity. The AFL-CIO and some five unions are now regular members of the forum.

A few years ago, only a rare labor leader would let himself be seen joining executives in an organization with "productivity" in its name. But given the current climate, unionists—like managers—cannot afford to overlook any possibility for strengthening their institutions for combat in the marketplace. As Glenn Watts of CWA observes, cooperating with management may appear to expose a union to high risks, but some form of cooperation is essential for the long-run survival of both parties. "The only real risk," he says, "is if the union does *not* participate."

August 24, 1981
Research associate: Faye Rice

11

A NEW SPIRIT IN ST. LOUIS

IRWIN ROSS

Union militance may be rising against the Reagan Administration, but when it comes to relations between unions and management, the mood around the country is anything but confrontational. In depressed industries such as auto, rubber, trucking, and airlines, organized labor has become surprisingly tractable. The auto union has made concession after concession to keep Chrysler afloat. Rubber workers and airline employees have voluntarily taken pay cuts. The teamsters union recently agreed to an early reopening of its contract with hard-hit long-haul truckers. And in a variety of industries, organized labor has begun to cooperate with management to improve product quality and compete more effectively with imports.

Probably nowhere has this cooperative spirit gone further or achieved happier results than in the St. Louis construction industry. Moreover, the cooperation in St. Louis has endured for nine years, long enough to prove that it is no fleeting burst of enthusiasm. The goals—to eliminate jurisdictional strikes, improve productivity, and make St. Louis an attractive place to build—have been realized to a remarkable degree.

Since 1972, there has been only one jurisdictional strike in the construction trades; previously three or four might occur in a month. More than three hundred restrictive work rules have been eliminated. Customers have been getting their buildings built on time. One agreeable result: commercial and industrial building is buoyant in both the city and county of St. Louis. And St. Louis construction remains a bastion of unionization—no mean achievement considering the immense growth in open-shop operations around the country. Nationally, more than 60 percent of all construction is now open shop, up from 30 percent in 1973.

St. Louis contractors report impressive gains in productivity. Gerald K. Sauder, former head of the mechanical contractors association, puts the improvement in the mechanical trades—plumbers, pipefitters, electricians, asbestos workers, sheet-metal workers—at 10 percent, and more for large projects because of the absence of jurisdictional strikes. James J. Murphy, Jr., a former head of the plumbing contractors association, also comes up with a 10 percent figure for the cost savings. Overall, Timothy R. McCarthy, president of McCarthy Bros. Construction Company, the largest general contractor at work on St. Louis projects, estimates productivity improvements for the combined crafts at somewhere between 10 and 15 percent.

This litany of success has been the handiwork of a union-management organization called PRIDE, an acronym for the estimable, if inelegant, slogan "Productivity and Responsibility Increase Development and Employment." PRIDE acts as a forum, gadfly, mediator, and policeman, solving disputes and spurring its members to more efficient operations.

Its results have helped inspire half a dozen similar organizations in other cities—among them MOST in Columbus, Top Notch in Indianapolis, Union Jack in Denver, and PEP in Beaumont, Texas (see box). PRIDE'S cochairmen, Alfred J. Fleischer and Richard Mantia, have been in great demand to speak before construction-industry groups eager to hear their secrets. The two men are zealots for union-management cooperation. Both are outgoing, ebullient, and nonstop talkers. Fleischer, the third-generation head of a family-owned general contracting business, is a conservative Republican who is a fervent admirer of trade unionism—in construction. "Organized labor gives us a trained cadre of men who are paid the same by me and by my competition," he points out. He praises their skill and likes the stability of the union arrangement.

Dick Mantia, executive secretary-treasurer of the local Building and Construction Trades Council, the umbrella organization for the building unions, was an asbestos worker for years before he politicked his way up the union hierarchy. He is a massive man who walks with a light step and sports diamond rings on two fingers—not an uncommon adornment among building craftsmen, who used to be known as "the aristocrats of labor." He calls himself "a liberal Democrat getting a little more moderate"—and he and Fleischer frequently argue amiably

---

SLOW CONTAGION FOR A GOOD IDEA

Joint labor-management efforts to increase productivity in construction are still uncommon, but at least ten local groups are devoted to that goal in addition to St. Louis's PRIDE. Most have adopted acronyms and logos for use in advertisements and on bumper stickers, placards, and construction signs. PRIDE influenced the formation of many of these groups, although it is not the oldest. Toledo's Joint Conference Board has been working quietly since 1939 to promote harmony in the fractious industry. The board's first objective was to reduce work stoppages caused by jurisdictional disputes between unions. Today stoppages are almost unknown.

Denver's Union Jack has been able to prevent nearly all walkouts due to jurisdictional disputes since its formation in 1976. The group has also recommended some 20 work-rule changes, more than half of which have been adopted. Among them are uniform shift hours and holidays for most crafts and a reduction in overtime pay from double time to time and a half for many.

In most cases, the stimulus to organization was a profound malaise in the local construction industry. In Indianapolis, the immediate cause was colossal unemployment in 1975—by some local estimates it was at least double the national average of 18 percent. Management and unions opened extended discussions, in the course of which they heard of PRIDE's success and got a copy of its memorandum of understanding. Indianapolis's Top Notch was founded that year. Its basic documents and methods closely resemble PRIDE's.

Top Notch has gone further than PRIDE in actively seeking construction work for Indianapolis; union management members have successfully lobbied for nearly $150 million of projects financed partly by the state or the combined city-county government. Top Notch's record in promoting labor peace has been outstanding: for five years there have been no strikes over economic issues and only one jurisdictional work stoppage, which Top Notch officials settled after two days.

Columbus's MOST (Management and Organized Labor Striving Together) was organized in 1976 as a result of apprehension about a large increase in open-shop activity in the area. Much of the problem resulted from delays and cost overruns caused by jurisdictional strikes. They have been eliminated on large construction jobs that have been designated MOST projects. To get more work, four unions—the carpenters, electricians, plumbers, and sheet-metal workers—have agreed to do residential work for 60 or 65 percent of the pay they get for heavier construction such as office buildings, factories, and government structures. One result is that unionized building in Columbus has increased from about a fifth to half of the total during the MOST years.

PEP (Planning Economic Progress) in Beaumont, Texas, is unusual in that it resulted from the initiative of Mayor Maurice Meyers, who took office in 1978. Appalled at the turbulent labor scene in Beaumont, a highly unionized oil-refining city 90 miles from Houston, Meyers looked at the techniques of PRIDE and promoted the same concept in his town. PEP not only includes all the construction unions, but the machinists and the Oil, Chemical and Atomic Workers, an industrial union that has occasional jurisdictional squabbles with the construction unions. In its first two years, PEP has helped settle 200 labor-labor and labor-management disputes. In pre-PEP days, these would have led to endless wrangles on the picket line.

about politics. He became interested in labor-management cooperation, he says, "because I could see the chaos that existed on the job sites. You didn't get your money's worth when you built in St. Louis." And he feared the encroachment of the open shop. "If we were going to keep the AFL-CIO here, we had to keep AFL-CIO contractors in business."

When it was adopted, the acronym PRIDE was a sadly ironic misnomer—aspiration rather than reality. "We probably had the most notorious labor reputation in the country," says John Kohnen, now president of the St. Louis chapter of the National Electrical Contractors Association. Carpenters recurrently quarreled with bricklayers, ironworkers with glaziers and millwrights, pipefitters with just about everybody else over work that they claimed employers had "misassigned" to a rival union. A jurisdictional conflict would generate a hot argument; the business agent would pull his men off the job and post pickets. Sometimes only one craft stopped work; sometimes the whole job shut down. In 1971, 102 days were lost in jurisdictional strikes affecting individual craft unions; thirteen days were lost in work stoppages affecting entire projects.

Prior to PRIDE, says Barry H. Beracha, a vice-president of Anheuser-Busch, it was realistic to expect two jurisdictional strikes on a twenty-four-month construction job, each lasting from a few days to three weeks. In 1970, when the St. Louis brewer embarked on a major modernization of its loading facilities, it obtained a project agreement, signed by all the unions, guaranteeing labor peace. "Two weeks later, it was broken by the pipefitters, who had a conflict with the ironworkers," Beracha recalls ruefully. The work was stalled for five months.

In the late 1960s the agonies of constructing two large projects—the Poplar Street Bridge across the Mississippi and the Gateway Ammunition Plant—scandalized both the construction industry and the town. Jurisdictional disputes, eruptions of violence, and walkouts for all manner of reasons resulted in endless delays and cost overruns.

One murder and any number of assaults occurred before the bridge was complete. At one point, after a carpenter was beaten, all the carpenters walked off the job to protest "terrorism" by the laborers. Some weeks later the laborers deserted their jobs, claiming that it was too cold for outdoor work. Without laborers to fetch their lumber, forty-one carpenters would not work; the following day they did not report. In the end, the contractor had to adopt ten- and twelve-hours shifts, enormously fattening everyone's wages with overtime pay.

From a public-relations point of view, what hurt St. Louis construction most was the noisome reputation of the pipefitters' Local 562, which ultimately attracted the attention of the national media. Lawrence L. Callanan, an ex-convict, took charge of the local in 1945 when it had fewer than three hundred members. Over the years he extended its sway across much of eastern Missouri, building it to some fifteen hundred members. He welcomed other ex-cons as members and used their muscle freely to cow other construction unions.

He was also not loath to shake down contractors and served five years in prison for extortion. Out on probation in 1960, he was under a court order not to resume the titular leadership of his local, but a commutation of his sentence by Lyndon Johnson enabled him to regain office in 1964. The local's political fund soon contributed $55,000 to national Democratic coffers, thereby creating another scandal.

Callanan died in bed in 1971, but his succession was marked by the same bloodshed that had intermittently characterized his regime. His successor, Edward L. Steska, was shot to death in his office. Thereafter, Callanan's son, Tom, a business agent in the union, lost his legs in a car bombing—a form of violence to which St. Louisans still seem peculiarly addicted.

Larry Callanan had made life tough for mechanical contractors, who employ pipefitters to install heating equipment and various kinds of industrial piping. For years there was an unwritten law that the union rather than the employer determined which men were to be laid off or transferred when total manpower needs on a job dwindled. Perhaps the most notorious work rule was a requirement that at least four men be hired to install pipe four to six inches in diameter, and at least six men for pipe eight inches or wider. The pretext was the pipe's weight, though there was no stipulation about its length.

The costly burdens of construction under such conditions disenchanted some of the biggest local building customers. Monsanto, the chemical giant headquartered in St. Louis, let it be known that it was going to do no more major building in the area. As for Anheuser-Busch, "We looked at construction projects in St. Louis with great caution," says Beracha. "Marginal projects were set aside."

By 1971 leaders of the St. Louis construction industry as well as some union leaders were becoming alarmed about the situation. After a building boom in the mid-1960s, work had declined sharply. In 1965 five of the "basic" construction trades—carpenters, cement masons, ironworkers, laborers, and operating engineers—had put in a total of 20.4 million hours on industrial and commercial work. The figure had fallen to about 13.6 million in 1970, with every indication that the decline would continue.

Seeking a solution, the contractors associations—more than a dozen in the highly fragmented industry—banded together to form the Council of Construction Employers. About the same time the St. Louis Construction Users Council was formed. It was a local offshoot of the newly formed Construction Users Anti-Inflation Roundtable headed by lawyer Roger Blough, the retired chief executive of U.S. Steel.

Its purpose was to try to arrest the steeply escalating cost of construction around the country, partly caused by outlandish union wage settlements and to some extent caused by hurry-up, "cost be damned" construction schedules set by a few huge companies. St. Louis was not leading the wage surge, but it had enough problems of its own.

The contractors and the users council took the lead. Early in 1972, says Al Fleischer, "we realized that if we didn't have the unions, we wouldn't have anything." Dick Mantia was an early enthusiast; so was Arthur A. Hunn, the veteran head of the painters union who was then president of the Building and Construction Trades Council. Unemployment in the building trades was around 30 percent. The members were restive, and clearly something had to be done.

PRIDE was formed on August 28, 1972, with no outside advisers. "We figured we were smart enough, cocky enough, arrogant enough to know our own problems," says Fleischer. From the outset, the group was broader than labor and management, also including representatives of the users council, architects, engineers, and, later, materials suppliers. The geographic area covered was larger than St. Louis. It embraced a substantial chunk of eastern Missouri, reaching out as far as the jurisdiction of most of the thirty-five locals in the building-trades council. Two-thirds of the cost of PRIDE is paid by the contractors and one-third by the unions. Recently it has amounted to a trifling $7000 a year; there is no paid staff.

A "memorandum of understanding," signed at PRIDE's launching, described each group's commitment. The unions pledged an end to jurisdictional strikes as well as "no limit on production by workmen nor restrictions on the use of tools or equipment." The promise to rid the industry of featherbedding practices was made more emphatic by the memorandum's pledge that "unnecessary and/or inefficient work practices, where they exist, shall be eliminated."

For their part, the contractors promised to get materials and equipment to the job site on time. They also, oddly enough, committed themselves to "exercise their management rights [in] directing, hiring, firing, and layoffs." In the past contractors frequently had abdicated these responsibilities to strong-willed union business agents. The customers pledged, among other things, to avoid unrealistic completion dates, which increased overtime costs. Architects and engineers pledged, in effect, to be efficient and accurate in their specifications. As Fleischer put it at at the inaugural luncheon: "Too often labor . . . has taken the blame for the expenditure of more man-hours than a project ought to have because of changes made by the design team."

PRIDE's organizers also created the machinery that makes the idea work—a board consisting of five representatives from the unions, five from the contractors, and one each from the users council, the architects, engineers, and, later, the suppliers. The board meets monthly—not merely when a crisis occurs—and discusses such concerns as future contract negotiations, jurisdictional problems, prospects for new construction, and proposals for special attention to large and prestigious construction jobs, on which the unions make renewed commitments not to strike. "All of a sudden we were talking regularly. We had a channel of communication," says Mantia. "Before PRIDE, we only saw each other at contract time when the talk could be pretty hot."

The transformation was not immediate. "The members needed educating," says Mantia. "At first they saw this as a giveback program." Mantia was in a good position to do the educating, for as staff head of the building-trades council, he was elected by a delegate assembly of the affiliated unions, not by the rank-and-file members; therefore he was in no danger of being voted out of office. He faced a lot of hostile questioning when he began to urge cooperation with the bosses, but he was not booed, he says.

RENOUNCING MAKE-WORK

In terms of restrictive work practices, the unions gave away a lot. The pipefitters were especially sweeping in their renunciation. Like the rest of the trades, they were suffering from unemployment. Taking a broader view than his predecessors of the causes of their plight, their new leader, Virgil Walsh—Callanan's second successor within a year—suddenly turned his back on the past. Among other things, he ended the linkage between the size of a work crew and the diameter of pipe to be installed, as well as any other specific manning requirements. He agreed to let contractors name their own foremen and to decide what workers to lay off and transfer. Harold Foley, the present pipefitters' leader, has been equally sympathetic to employers' beefs.

Other unions also relaxed onerous restrictions. For years, home-builders had been required to erect a tool-storage shed on each job site. This was eliminated when parking was available next to the site, which enabled the workmen to store their tools in their cars. An operating engineer is now allowed to run three small machines, such as compressors, pumps, and welding machines. Previously, an engineer could operate only two machines at a time, with a second man required if a third machine were to be used. A foreman, a union member, used to be required for every six laborers on the job. The ratio is now one to ten.

Seemingly small changes generated substantial cash savings for contractors. Such items included flexible starting time, flexible lunch time, and the reduction or elimination of paid travel time from St. Louis when jobs in the hinterland were close to the workers' homes.

The immediate end to the epidemic of jurisdictional strikes was an even more dramatic consequence of PRIDE, marred by that one lapse in nine years. One of the most helpful techniques toward job-site peace has been the use of prejob conferences on projects of any substantial size. In attendance are a representative of the owner, the general contractor, each subcontractor, business agents from the crafts likely to work on the job, and Dick Mantia. An effort is made to determine in advance, for example, whether glaziers or ironworkers will install certain windows (it depends on the kind of frame used) or whether carpenters or bricklayers affix Styrofoam insulation to walls. (If the stuff is nailed, it's carpenters' work; if it is applied with adhesive, it is bricklayers'.) The prejob conferences avoid many "misassignments," which are difficult to change once made.

Conflicts still occur on job sites, of course, but they are resolved through negotiations. First the business agents of the disputing locals confer. If they cannot agree, they call in representatives of their respective internationals. One side or the other generally backs down—the same thing that occurred in the old days, of course, but often after a walkout. Until recently, unresolved matters would also occasionally be submitted to the Impartial Jurisdictional Disputes Board in Washington, D.C., a joint union-management enterprise that suspended operations earlier this year. Its drawback was that it took four to six weeks to make a decision.

On occasion Fleischer and Mantia are rushed in as an emergency squad to prevent a dispute from boiling over. In 1975, when the city's convention center was under construction, an out-of-town subcontractor was employed to install forty-foot-high metal doors. He hired ironworkers to do the job—much to the horror of the carpenters' union leader, Ollie Langhorst, who argued that in St. Louis this had always been a job for millwrights, who are in the carpenters union. Langhorst was right about the precedent, but Fleischer couldn't persuade either the general contractor or the subcontractor to change the crew. Meantime, the ironworkers were already installing overhead tracks for the doors. Fleischer and Mantia had several exhausting sessions with Langhorst. They finally struck a compromise, in which the ironworkers completed the track installation and millwrights hung the doors. Before PRIDE, the carpenters might have hit the bricks, and the job probably would have been shut down.

Strikes at contract-expiration time are not outlawed by the PRIDE memorandum, of course, and five serious ones lasting at least three weeks have occurred. The unions have agreed, however, not to picket during strikes, thereby allowing other crafts to continue work. Fleischer and Mantia invariably offer their services as mediators. In July 1981 they worked out a face-saving compromise that ended a twenty-four-day bricklayers' strike.

One impressive piece of evidence of PRIDE's effectiveness is that on-time completion of large construction projects, without cost overruns, is now commonplace in St. Louis. The latest example is the $50-million First National Bank Building, a dazzling glass tower that is the newest addition to the city's skyline.

Several buildings have been finished under budget and ahead of schedule. The Blue Cross Center opened five months early and $1 million below its budget of $12 million. The city's prized convention center, expected to take two and a half years, was completed three months sooner than that and cost $864,000 less than the $35 million projected. Hotelier Don Breckenridge, who put up what is now the Marriott Pavilion, beat his schedule by two months and spent 4 percent less than he had planned. Breckenridge is currently building a luxury hotel at the airport for $28 million, or

$71,000 a room. He insists that a comparable hotel would cost $80,000 to $85,000 a room in most other places in the country. The difference is mainly St. Louis's increased productivity.

Major companies that were reluctant to build in St. Louis in the early 1970s—Monsanto, Procter & Gamble, and Anheuser-Busch, among others—have been doing considerable construction during the PRIDE years. The area is enjoying a building boom, with $2.3 billion in construction under way in the Missouri portion of the St. Louis metropolitan area.

Dick Mantia estimates the unemployment rate among his flock at 6 to 7 percent, compared with 30 percent when PRIDE started. If residential construction were not so depressed, he insists that there would be a shortage of labor. "Meantime," he boasts, "not one of our contractors has gone open shop." They have had no need to.

November 16, 1981
Research associate: Elizabeth S. Silverman

12

WHY GOVERNMENT WORKS DUMB

JEREMY MAIN

Politics and productivity don't seem to mix. Politicians denounce waste and mismanagement but, once elected, usually appear incapable of improving government effectiveness. It's a pity, because the increases in productivity achieved by Japanese manufacturers and by the best-managed U.S. companies, if applied to government, would deliver huge savings to the taxpayer. The Joint Economic Committee of Congress figured in 1979 that a 10 percent improvement in the output of the nation's 2.8 million federal workers would lop $8 billion out of the budget—without reducing services.

Like Proposition 13 in California, Ronald Reagan's unprecedented budget cuts will prod civil servants to do more with less. But wholesale reductions alone are not enough and sometimes hurt productivity. To achieve short-run economies, government officials often eliminate the capital investments, maintenance, planning, and training that would improve performance in the long run. In its fiscal crisis of 1975, for instance, New York City chopped deeply into maintenance and capital spending, accelerating the deterioration of its buildings and equipment that has sapped services and productivity.

Reagan's knife is slicing into the Office of Personnel Management's efforts to organize a federal productivity drive—meager as those efforts have been in the past. As business has discovered, productivity campaigns aren't likely to catch on without a firm commitment from the chief executive. Although a few state and local governments have put together successful productivity programs, Reagan, like his predecessors, has not yet shown interest in adopting a comprehensive system that sets goals, measures results, and rewards performance.

Here and there—in Phoenix, in North Carolina, in some federal agencies—significant gains have been made. But most political entities haven't gone beyond lip service. The lack of progress is partly concealed by a great show of effort across the country. Productivity has become a buzzword among bureaucrats. The output of reports on the subject is itself an awesome feat of productivity.

But the political system is stacked against productivity. By nature, headway in improving performance is made in small steps over a long time. Politicians, however, need quick, highly visible accomplishments that will help them win the next election—and sexy issues that can be covered in a thirty-second TV spot. "We've got a long way to go before productivity has sex appeal," says North Carolina's two-term governor, James B. Hunt, Jr., one of the few elected officials who have launched serious campaigns to improve efficiency. "People are suspicious," he says. "All they've heard about it is bad. You've got to show them you can maintain services while absorbing budget reductions." Hunt doesn't try to talk about productivity to most audiences for fear of putting them to sleep.

Improving productivity in government is far more difficult than in business. Government has no unambiguous bottom line. The rewards go to those who expand their staffs and win big appropriations, not to those who find ways of saving money. "Managers in government know that the good guys get hurt," says congresswoman Patricia Schroeder, Democrat of Colorado, who is pushing a bill to force agencies to lay out clear objectives and be judged by how well they meet them. "If you're lean, mean, and effective," she says, "you get killed." Mrs. Schroeder argues that "a productivity program need not be expensive. Canada has, over the past seven years, instituted what appears to be a first-rate program at a cost of $13 million. That's million, not billion. Around Washington, $13 million is mad money."

AN ACRE OF IDLERS

The public thinks civil servants do little and do it badly. The most recent Harris Poll on the subject (in 1973) rated government employees' output far below that of other workers. Anecdotal evidence of sloth abounds. The Agriculture Department is famous as a citadel of idle talk and unoccupied offices. A productivity consultant who attended a conference a few years ago in the General Accounting Office sat for three days at a vantage point where he could watch about an acre of civil servants at their desks. He put his industrial-engineering talents to work tabulating the activities of people he could see. He concluded they were working only 20 percent of the time.

The Bureau of Labor Statistics, which measures government productivity, has produced figures that seem to contradict public perceptions. Every year the BLS surveys the hours worked in over three thousand government tasks, from delivering letters to locating deportable aliens. The survey now covers 373 agencies and departments—66 percent of the civilian federal work force. Since 1967, the BLS index shows, productivity for the federal government as a whole has been growing by 1.4 percent annually (see chart). In recent years federal productivity has speeded up while the private sector's has slowed, so that the government is now beating business in the productivity game.

Unfortunately, the measurements so laboriously collected by the BLS are far from exact. Gauging government activity is innately difficult. How do you measure the effectiveness of the military without a war? The BLS figures are published in categories so broad as to be of little use. Some agencies can obtain unpublished breakdowns showing how well they stack up against other parts of the government. But comparisons are often thwarted by the sparsity of statistics submitted to the BLS. For instance, the Veterans Administration lumps all its hospitals' figures together, so no one can compare them.

PAY OUTRACES PRODUCTIVITY IN WASHINGTON

Of all areas of the federal government, only communications achieved a decline in unit labor costs from 1967 to 1979. Automated equipment raised the productivity of communications employees by 9% a year, which more than offset pay increases of 6.9%, reducing unit labor costs by 1.9%. In all other categories, productivity gains failed to keep pace with compensation. The categories below are among 28 measured by the Bureau of Labor Statistics. Overall, federal productivity grew 1.4% a year, compared with 1.6% in the private sector.

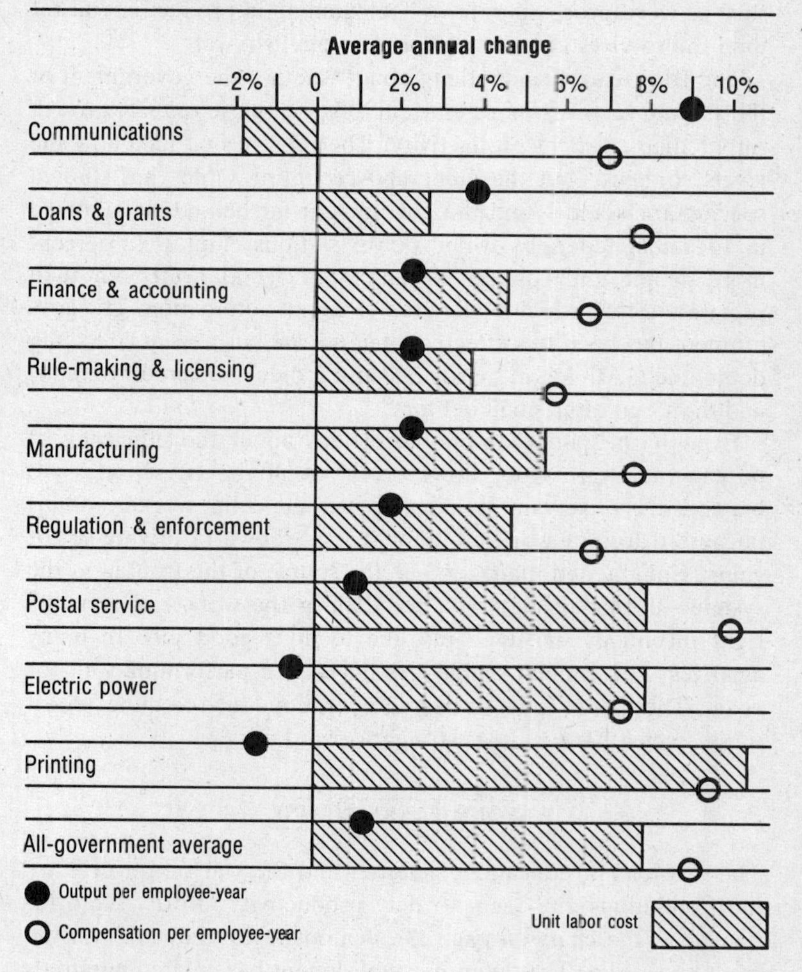

Even the broad judgment that government productivity has been climbing may be misleading. The BLS admits that productivity might actually show a decline if only net output is considered—that is, if the delivery of services by one government agency to another were eliminated. The agency believes that intragovernment activities, involving many automated operations such as word processing, have been gaining in productivity faster than the services actually rendered to the taxpayer.

The BLS figures say nothing about whether the government or the private sector is more efficient. They disclose *rates* of growth rather than *levels* of productivity. The BLS has no idea how the levels compare, but the General Accounting Office has studied specific areas and found the government far behind. It reported in 1979 that federally owned power stations employ 48 percent more people and cost 20 percent more per kilowatt-hour than private utilities, which make greater use of automation. The government has been less effective than private business at collecting debts, the GAO found, partly because it didn't use credit bureaus and didn't go after small debtors.

Even if the public's worst suspicions about the efficiency of bureaucrats were true, however, it would be as senseless to blame the civil servant as to blame the blue-collar worker and his purported loss of work ethic for the slowdown in private-sector gains. Enlightened managers see the source of the trouble in the system—in management—rather than in the worker. Given the right incentives, most people like to do a good job. In many agencies, government employees have been performing well for years. They seem to work best in operations such as the passport office, which have a clear, straightforward mission.

FOUR TANKS TO ONE

The missions of education, defense, and the war on poverty are more nebulous and seem to defy productivity efforts. Unfortunately, as the chart on page 146 demonstrates, they are also the most expensive. The defense establishment has created hundreds of admirable productivity programs. The Navy, for instance, has introduced quality circles in shipyards, pays bonuses to computer operators who raise their output, and has streamlined purchasing for "fast payback" equipment. But these efforts are drowned in a proliferation of checks, hearings, analyses, reviews, and other complicatons. Anthony R. Battista, a member of the House Armed Services Committee staff, figures it took thirteen steps to develop a new weapons system until Defense Secretary Robert McNamara began converting the Pentagon to systems analysis in 1961. Today it takes four hundred steps. The Russians progressed through four generations of tanks while the U S. developed one.

Few people in Washington would disagree with James M. Peirce, president of the National Federation of Federal Employees, who says morale is at "rock bottom." How can government workers be expected to produce more, he argues, when "their bosses [especially in the current Administration] and the news media have continually maligned, criticized, and misrepresented federal employees." Senior civil servants are leaving government at an unprecedented rate—20 percent planned to go between 1981 and 1983—mostly because their pay has been squeezed against a ceiling of $50,112*; the top seven grades are bunched together at this limit. In the last five years, top officials have had one 5.5 percent cost-of-living increase, while the consumer price index has gone up 44 percent.

President Reagan has swept like an avenging angel into this dispirited army, wielding his cost-cutting sword. While wholesale budget cuts may spur some officials to innovation, the bureaucracy has an extraordinary ability to deflect reform. A common reaction is known in the capital as the Washington Monument syndrome. Threatened with a reduction of funding some years ago, the National Park Service warned that it would have to close down the monument, its most popular facility. Naturally the monument was not closed, and, as expected, some funds were restored.

Even the most promising remedies have a way of going astray in Washington. In 1978 Congress passed the Civil Service Reform

*In January 1982, the limit was raised to $57,500.

THE BIG ONES ARE THE TOUGH ONES

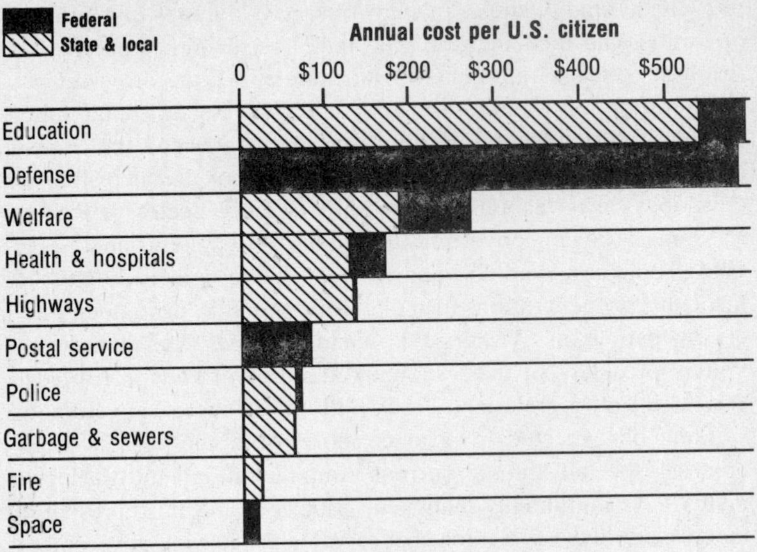

Federal
State & local

Annual cost per U.S. citizen

0 $100 $200 $300 $400 $500

Education
Defense
Welfare
Health & hospitals
Highways
Postal service
Police
Garbage & sewers
Fire
Space

Source: Bureau of the Census

The juiciest targets for productivity marksmen are the activities that cost most, where small gains could mean big savings. Unhappily, education, defense, and welfare are among the toughest government services to improve, while less expensive sanitation and fire departments are more easily streamlined.

Act to increase work incentives. The act created a Senior Executive Service of 7200 top positions in which officials can earn bonuses of up to $20,000 a year. The results, says an official of the Office of Management and Budget, "have been very disappointing." Instead of being used to reward competence—which in any case is hard to discern in government—the bonuses are often passed out in lieu of raises. With Reagan's apparent lack of interest, Washington doesn't seem to be getting any closer to an effective policy on productivity.

In contrast, North Carolina has a productivity program directed from the top. After isolated efforts in the early 1970s, the state government got pointed in the right direction with the election of Jim Hunt as governor in 1976. Now that he is in his second term, productivity is becoming more than a policy that can easily be reversed by the next election: it is becoming institutionalized.

The state's department of administration has put together a productivity staff, and each department has installed a senior official responsible for productivity. The governor has a permanent commission of business, state, local, and educational leaders to recommend new policies. About two hundred fifty state officials—most of the top three tiers of government—have taken courses designed for them by the graduate business school at the University of North Carolina. The government offers productivity-consulting services to its agencies. Incentive awards have begun to spread: individuals can win up to $5000 for suggestions, while groups that volunteer to achieve higher productivity goals can keep one-quarter of the savings.

The Rowan County road-maintenance crew became the first to sign up for North Carolina's incentive awards in 1978. Through small changes, such as using a new leveling procedure that reduces the base material needed on dirt roads, the unit managed to keep its expenses some 10 percent below the budgeted $600,000. When their share of the savings was divvied up among full-time employees, each got $243.95. The amount isn't vast, but it seems enough to interest employees at twenty-six other units that have signed up to work under the same system. The department has reduced its work force of 14,000 by 1126, mostly by attrition.

The North Carolina county land-records offices have also achieved striking productivity improvements. Typically, the records in Orange County, which encompasses Chapel Hill, were spread out among five offices using everything from computers to ledgers and pens. A sale of property would show up at the registrar of deeds, but a transfer by inheritance would be recorded by the clerk of the court. A lawyer's aide would spend a full day searching a title for an ordinary house. Now all the records related to taxes, liens, titles, and assessments have been filed in one computer memory that can be thrown onto a screen in seconds. The title-search time has been cut to two hours or less, which represents pure profit to the lawyers, since they haven't lowered their fees. The registrar of deeds has reduced its staff from six clerks to one, and other counties are learning how to do the same.

Phoenix has one of the best governmental productivity programs in the country. The city used to be a politically disorderly place that fired its manager about once a year. Then, in 1970, Phoenix called in Booz Allen & Hamilton, the consulting firm, which urged it to focus on productivity. At first, the city concentrated on the technical and industrial-engineering improvements that come easier than dealing with the performance of people. The police computerized their dispatch system, and three-man garbage trucks were replaced with one-man trucks, which are saving around $3 million a year.

In the mid-1970s, when Marvin A. Andrews, a productivity enthusiast, became city manager, he shifted attention to the human side. Management-by-objective and merit raises, unthinkable until recently for civil servants, were applied to the top two hundred officals. Anyone who meets mutually chosen objectives can earn a raise of up to 20 percent. The city's budget office estimates that annual spending would be 10 percent higher than the current $350 million were it not for the productivity drive.

Phoenix hires private contractors to perform many municipal functions. They collect one-third of the garbage and handle other activities, from landscaping to water-bill processing to services for senior citizens. Contractors often do a better job for less money than municipal workers, and many cities are using more of them. E. S. Savas, an assistant secretary at the Department of Housing and Urban Development and former Columbia University expert on city government, says his studies show that with better management and equipment, contractors can provide the same services for up to one-third less than municipal employees.

COMPETITION FOR THE GARBAGE

That doesn't mean private enterprise has the whole answer. Phoenix seems to have adopted the best stance by letting municipal and private organizations bid against each other; splitting garbage routes between private and public collectors keeps both on their toes. Budget director Charles E. Hill says, "There's a general openness about it—our people say, 'If the outside contractors can do it cheaper, let 'em.' The city also bids, and of course, if the city is cheaper we continue in-house." Phoenix ties contracts to quality controls and has fired contractors in midterm for nonperformance.

Boston University's controversial president, John R. Silber, recently offered to take over the Boston public schools and run them for $40 million less a year than the $245 million now being spent. The schools are a pathetic example of the conflict between productivity and politics. In the last decade the student population dropped by a third, but the elected school board increased the teaching staff by 13 percent and the nonteaching staff by 150 percent. Silber says B.U. has the managerial, budgetary, and academic skills the schools lack. "They don't have even an ancient budget system," he snorts. "They don't even know what they spend and how much they have left from month to month."

Financially pressed old cities such as Boston find it harder to achieve breakthroughs in productivity than prosperous young cities like Phoenix. New York is still handicapped by inefficiencies designed into the subway system at the turn of the century, when labor was cheap. One man can operate a modern subway train; New York's require two.

Having watched municipal services deteriorate during six years of fiscal turmoil, New Yorkers may be surprised to learn that their city's productivity program is highly regarded by administrations elsewhere. The trouble is that the program is long on reports about methods improvement, integrated financial reporting, and intervention models, and short on results that the man in the street can grasp. New York has replaced policemen on traffic duty with lower-paid agents and reduced some garbage-truck crews from three men to two. But all the efforts have hardly affected poor work habits, patronage, deferred maintenance, and other city ailments.

San Francisco, with a government almost as Byzantine as New

York's, woke up to its productivity problems late, but then decided on nothing less than a complete shake-up in the way government is managed and financed. The late Mayor George Moscone recruited as deputy mayor a C.P.A. and business professor, Rudy Nothenberg, who believes that "just because you're a liberal Democrat doesn't mean you have to be a fiscal slob." Nothenberg recalls, "We had to start from nothing. We began with no system. We couldn't define output. There had been no attempt to define management. We didn't know who the managers were."

The books were so muddled that San Francisco couldn't put out timely financial statements. In 1978, the year Dianne Feinstein became mayor after Moscone's murder, productivity got a strong push from Proposition 13, which threatened the city's budget. With considerable help from a business-backed Fiscal Advisory Committee and other private organizations, San Francisco trained its managers, automated and consolidated its bookkeeping, and implemented a new budgeting system. Rather than rely solely on a "line" budget, which apportions money in categories such as salaries and supplies, the city has adopted a "functional" budget that breaks down costs by individual service—fire prevention, maintaining Golden Gate Park, and so forth. City managers now negotiate objectives and are judged on how well they achieve them. Mayor Feinstein reviews the performance of department heads every quarter.

The city is also instituting an imaginative incentive program. Officials who join the Senior Management Service will be eligible for above-average raises if they perform well and no increases if they don't. Instead of managers just percolating up through the ranks over the years, they will now be able to ride a fast track—but be subject to demotion or dismissal. San Francisco is treating its managers in a most uncivil-service-like manner.

Gains are beginning to pile up, small and large. The new accounting system freed $30 million that had been tied up unnecessarily—and helped convince many old hands in the controller's office that reform wasn't all bad. Building permits are issued faster, police response time has been cut, and checks coming in to pay for property taxes are deposited in three rather than fifteen days, yielding an extra $300,000 a year in interest. The Fiscal Advisory Committee figures the city is already saving about $43 million in an annual budget of $1.2 billion. "A good deal of the impetus for our productivity drive came from Proposition 13," says Mayor Feinstein. "People are finding they can do certain things with less money. That has created a major boost in morale in government."

San Francisco, Phoenix, and North Carolina are among the exceptions. The great majority of states and municipalities have no productivity programs at all, not even the kind of leaderless experimentation seen in the federal government.

Judging by what is (and isn't) working around the country, the elements of a successful program seem fairly clear. When management-by-objective is laid over a functional budget, as in San Francisco, taxpayers can see if they're getting what they're paying for, and officials can be held accountable. Giving government managers specific goals is a wise, if somewhat novel, idea. Achievement can be further stimulated by bonuses, faster promotions, passing out a share of savings, or all three.

With these reforms and others, politicians can alter bureaucratic incentives and behavior. But unless they lend the task their committed long-term leadership, efficiency in public work will be nothing more than a passing slogan.

August 10, 1981
Research associate: Anna Cifelli

13

THE BATTLE FOR QUALITY

JEREMY MAIN

Why are we suddenly so worried about the quality of American products? Boeing makes the best commercial aircraft in the world. International Harvester and Deere & Company produce the most reliable tractors—equipped, if farmers want, with stereo, air conditioning, posturepedic seats, and adjustable steering wheels. European tourists carry off American-made permanent-press sheets because they're cheap, long-wearing, and dazzlingly patterned. Our plastics are stronger, our chemicals purer, and our machine tools built to finer tolerances than ever. This is all true enough—yet, unhappily, it is also largely beside the point.

America's leadership in quality has been almost imperceptibly eroding for years. More and more U.S. executives have awakened lately to the fact that they are caught in a fateful struggle. They are turning their companies upside down to give quality specialists more clout. Vendors are being told to supply better parts or lose the business. In hundreds of factories small groups of workers are sitting down periodically to search for ways of improving quality and productivity. Executive offices and factory floors ring with slogans. Says John A. Manoogian, executive director of product assurance at Ford Motor Company: "If it's not right, we won't ship it, and we mean that." Intones a General Electric spokesman: "Quality is our number-one focus."

While U.S. companies have steadily improved quality, they are coming under pressure largely because the Japanese have advanced by leaps. Building assiduously for thirty years on a foundation of theories developed in the U.S., the Japanese have made quality the weapon that wins the world's markets. A few bald facts show how well they have succeeded: A new American car is about twice as likely to have a problem as a Japanese model. An American color TV needs repairs half again as often as a Japanese set. U.S.-made computer-memory chips were judged in a test in 1980 to be three times as likely to fail as Japanese chips. "There's no question that the Japanese have set new world standards," says Robert E. Cole, who has worked in Japanese factories and now directs the University of Michigan's Center for Japanese Studies. "Their best factories are better than our best factories."

Robert B. Reich, director of policy planning at the Federal Trade Commission, forcefully summarized the consequences of the relative decline in American quality in a speech in 1980: "In industry after industry, consumers in America and elsewhere are turning their backs on U.S.-manufactured products in favor of foreign competitors; twenty-eight percent of our automobiles are now manufactured abroad by non-U.S. companies, thirty percent of our sport and athletic goods, thirty-four percent of microwave ovens, ninty percent of CB radios and motorcycles, almost one hundred percent of video cassette recorders. The list goes on and gets longer year by year: radial tires, calculators, televisions, food processors, premium beer, cameras, stereo components, digital watches, pianos, bicycles, outboard motors."

Among U.S. companies, it's often hard to identify the special characteristics that divide those that lead the league in quality from those that have fallen behind. Certainly the pacesetters are run by people who insist on excellence, often by an individual or family with a reputation to protect. Beyond that, though, none of the usual influences on industrial performance—unionization, capital intensity, the number of competitors—seems to explain the quality disparities among companies or industries. Nor is there reason to suspect that today's makers of even the best U.S.

products—farm equipment, aircraft, machine tools, large appliances—will be immune to future threats. One thing that complicates thinking about quality is that the word can mean just about whatever a customer thinks it means. To the fashion-conscious, quality might mean the patch on the seat that transforms a $20 pair of jeans into a $40 pair. To a space scientist, quality represents a million parts so carefully made, tested, and assembled that they will function flawlessly for years. To the Department of Agriculture, quality in food means uniformity and an absence of "gross contamination and adulteration"; to a diner at New York's Lutèce, it may mean the crayfish just flown in from New Orleans or a tart of the airiest pastry smothered with fresh raspberries.

You might think of quality as craftsmanship—a seamstress handstitching an evening gown or a stonecutter carving a gargoyle. There is not much of that left. At issue today is the quality of manufactured goods, which need little craftsmanship but are frequently more reliable and durable than their handmade precursors. A working definition of the quality most sought today might be "fitness for use, plus reliability, delivered at a marketable price."

The plainest example of the potential for interminable argument about quality is probably the automobile, a leading subject of the quality alarm. Detroit admits that Japanese cars have better-fitting doors and neater paint jobs—but argues that American "fits and finishes" are improving. Volkswagens and Toyotas win on fuel economy, although American cars are improving here too. When it comes to frequency of repair, J.D. Power & Associates, marketing consultants in Los Angeles, recently found that 64 percent of the owners of five types of new American cars reported mechanical difficulties in the first six months, vs. 35 percent for Japanese cars. On the basis of reader surveys, *Consumer Reports* says that Toyotas, Datsuns, Hondas, Mazdas, Subarus, Volkswagen Rabbits, and Mercedeses need repairs far less frequently than other cars.

In other respects—for instance, comfort, quiet, corrosion resistance, reliability of engine and transmission—American cars can claim the advantage. Japanese cars fare poorly in safety tests. In

August 1980 the National Highway Traffic Safety Administration flunked all the Japanese entries in a crash-barrier test; three American cars passed. On balance, you might argue that American cars are now as good as the imports, but the public clearly doesn't think so. In a recent survey for the American Society for Quality Control, Americans panned U.S.-made autos. That judgment may be too harsh, or lag behind the facts, but perception can be as important as reality. It has the same effect in the marketplace.

However quality is defined, the American consumer is clearly ready for more of it. Since the early 1970s Yankelovich, Skelly & White, an opinion-research firm, has been keeping track of what people think of the products they buy. Every year a greater proportion of those surveyed say they would be willing to pay more for better quality.

The principles for improving quality were developed in the U.S. before and during World War II—and then neglected here. But Americans have been teaching the principles of excellence to the Japanese for three decades, with extraordinary results. When industrialists moved into the postwar power vacuum in Japan, they saw that the economy depended on wiping out the nation's image as the world's purveyor of junk. In 1950 the Japanese invited W. Edwards Deming, a Census Bureau statistician, to lecture on methods he had developed for statistical analysis of quality. Four years later they asked Joseph M. Juran, who was born to a village shoemaker in the Balkans and became a Western Electric quality manager in 1935, to instruct them on management's role in improving quality.

While both men had been pretty much ignored in the U.S.—the attitude was, "Go away, Deming, we're making money," according to a colleague—the Japanese honored these prophets by devotion to their teachings. The Deming prize for quality control, now in its thirtieth year, became so important in Japan that the annual award is broadcast on national TV. Now, finally, American businessmen are importuning Deming and Juran for advice. Juran, a crusty 75-year-old, says, "I've never been busier in my life." Imperious at 80, Deming takes calls from Fortune

500 companies between trips to Japan and Europe. "They want me to come and spend a day doing for them what I did for Japan," he snorts. "They think it is that simple. American management has no idea what quality control is and how to achieve it."

In fact, the statistical controls that Deming taught the Japanese are deceptively simple, at least in their essentials. You tally defects, analyze them, trace them to the source, make corrections, and then keep a record of what happens afterward. A classic case, described by an associate of Deming's, Professor David Chambers of the University of Tennessee, illustrates the principle. In the 1960s the managers of a middle-sized hosiery plant in Tennessee, now owned by Genesco, decided they had to change the ways they had grown used to in the previous sixty-five years.

A preliminary study showed one source of defects was the "looping" department, which performed the task of closing toes. Inspectors soon found that a few of the workers were responsible for most of the defects. An older worker responsible for 20 percent of the faulty loops was persuaded to take early retirement. A dozen others did fine once they got new glasses. Another said she paid little attention to what she was doing because no one had shown concern for quality; when management did, she easily improved her work. So it went. In seven months, the plant cut its rejects from 11,500 to 2000 stockings out of a total weekly output of 120,000 pairs. To put it another way, productivity climbed 4 percent virtually cost-free. Until recently few American companies bothered with that kind of analysis.

But a few fixes on the shop floor are rarely enough. Effective quality control—or "quality assurance," in the phrase favored by specialists today—involves the whole industrial process, beginning with product design and continuing on to the marketplace. In the past, American companies usually controlled quality by putting enough inspectors in the plant to weed defects down to an acceptable level. To people like Deming and Juran, you have already failed if you need a lot of inspectors. But the Japanese approach is taking hold here. "You can't inspect quality in," says Alex Mair, head of GM's technical staff. "You must build it in."

In the past, quality experts looked at a design after it was finished. If they didn't like what they saw, they had a bureaucratic fight on their hands. But when they are involved from the beginning, they can make sure the product is reliable and easy to make, which should lead to fewer defects. Mair says GM's X cars were the first to benefit from quality management starting from the design stage and as a consequence were introduced in 1979 with fewer bugs than usual. "We made a giant step with the X cars," says Mair, "and by the middle of the decade the process will have had a noticeable effect on all our products." *

The normal vendor-client relationship in the U.S. is adversarial and arm's-length. The client sets minumum standards and rejects items he considers defective. In Japan the vendor is brought into the whole quality-improvement program. When they set up factories here, Japanese and German companies are often surprised to find that vendors cannot meet their standards. But they don't accept the status quo. Volkswagen not only insisted on higher standards but required vendors to reorganize their managements to give quality control higher status.

The U.S. auto industry has lately adopted a new attitude toward vendors that combines tougher standards with a willingness to cooperate in solving problems. Beginning with the 1980 models, Chrysler started requiring executives to make sure vendors have the tools, manpower, and whatever else is needed to produce parts at the specified level of quality and volume. Unless they are satisified, the Chrysler representatives won't approve a supply contract. The program covered 80 parts for the 1980 models and 516 for the 1981s.

RUNNING AROUND THE CLOCK

Good quality demands good tools. A $100,000 machine that can weld perfect seams all day and night, seven days a week, without

*Mair seems to have been prematurely optimistic. In answer to a survey, *Consumer Reports* readers said after a year's experience that the X cars had to be repaired more often than the average car. The X cars were also recalled seven times in the first year because of safety problems.

ever taking a break or complaining about the heat and noise, can raise both productivity and quality. Simple automatic tools have been around for a long time, but computer-controlled robots that can be reprogrammed to do many tasks, that can "see" and "feel," that can inspect their own output, that can adjust for wear on their working surfaces, are only beginning to appear in American factories in large numbers. The five-year, $75-billion retooling of the U.S. auto industry has created a surge in demand.

Robots are used principally for welding auto bodies, a dirty, tiring job that humans tend to flub at the end of a shift. They are also good at inspecting, because their attention never flags. TRW plans to install robots that use ultraviolet light to look for flaws in fan blades for jet engines. That requires perfect inspection, for a faulty blade could cost hundreds of lives.

Cincinnati Milacron, Inc., a major maker of machine tools and robots, predicts demand for U.S.-made robots will grow by 25 to 30 percent a year for the rest of the decade. About twelve hundred robots were produced in the U.S. in 1980. James A. D. Geier, grandson of Milacron's founder and now its president, sees quality improvement coming as robots spread through industry. "They run round the clock, mostly unsupervised, turning out parts that always fit, last longer, and run better," he says.

In the popular imagination, the loss of the "work ethic" in America is to blame for poor quality, but specialists disagree. Under proper management, the U.S. worker is as good as anyone. In fact, some of the best German and Japanese plants are in the U.S. The Volkswagen Rabbits built in Pennsylvania and the Mercedes trucks assembled in Virginia meet German standards. Sony's San Diego plant set a company-wide record by turning out Trinitron TV sets for two hundred consecutive days without inspectors finding a single operating defect.

The turnaround at a former Motorola factory in Franklin Park, Illinois, has become a famous example of what the U.S. worker can do. Under Motorola, the plant was literally turning out more defects than TV sets: inspectors found one hundred and forty defects for every one hundred TV sets that passed along the lines. Matsushita Electric Industrial Company bought the plant in

1974 to make Quasar and Panasonic TVs. The Japanese kept the same labor force and even chose Motorola's vice-president of engineering, Richard A. Kraft, to be president of the new subsidiary. Kraft says the rate of rejects is now down to four to six per one hundred sets, and the number of warranty claims has been cut eightfold.

The new management made wholesale changes, large and small, on the plant floor. Matsushita repainted the walls and put the workers' names up at their stations. Automation was increased, leaving less chance for human error. Most of the moving belts on the production lines were replaced with a system the workers can control. If an employee sees a problem, she (most are women) can take the time needed to solve it. She still has to meet quotas.

To draw workers into the process of improving quality, corporations are turning their backs on the old style of "scientific management" that grew out of Frederick Taylor's theories early in the century. It's now recognized that quality suffers when employees are assigned to isolated, limited tasks and don't understand what they are doing. Unfeeling managers and uncaring workers often become adversaries despite mutual interests. Today workers are being asked to think about what they are doing and how it can be done better. The quality circle is a popular vehicle for involving blue-collar employees in quality and other improvements. In Japan about five million workers belong to circles of about ten people each. The circles made their mark in the U.S. early in the 1970s, when GM used them to help turn around a badly demoralized plant at Tarrytown, New York. Quality improved markedly, while absenteeism and grievance claims dropped. About eighty-five GM plants are now experimenting with circles. Ford and Chrysler are working on similar ideas, with the enthusiastic cooperation of the United Auto Workers.

Some two hundred U.S. companies are trying circles, though not always successfully. Unless the workers get some training and help from a quality specialist, the circles can become aimless kaffeeklatsches. But Cincinnati Milacron is pleased with its twelve circles and plans more. "We are seeing enormous results in better quality and better processes," says Geier. The results range from supplying one department with more brooms to designing a gauge that reduced the rejection rate on a hard-to-make shaft from 50 percent to zero.*

Bringing design and quality engineers, blue-collar workers, vendors, and sales and service forces into a cooperative effort to improve quality is management's responsibility. Deming holds that only 15 percent of all quality problems—being a statistician, he calls them variations—are related to a particular worker or tool. The other 85 percent arise from faults in the company's system and will continue until that system is changed. When people tell him, "Ed, I'm getting another three-million-dollar machine," Deming says he replies, "What you need around your place is three hundred thousand dollars' worth of brains, not another machine."

SUPPORTED BLASPHEMY

The contrasting attitudes of American and Japanese management offer perhaps the clearest explanation for the different quality of what they produce. Stephen Moss, an Arthur D. Little consultant who has worked with corporations in both countries, says, "The U.S. manager sets an acceptable level of quality and then sticks to it. The Japanese are constantly upgrading their goals." The American assumes a certain rate of failure is inevitable, adds Moss, while the Japanese shoots for perfection and sometimes gets close.

Japanese culture helps instill the right attitudes in management. A Japanese executive expects to spend a lifetime with one company, so its long-term success is his success. An American takes a more self-centered view of his career. American business can hardly import Japanese culture wholesale, but the U.S. is not doomed to being second-rate. Says E. James Stavrakas, who runs J.C. Penney's quality-testing operation: "There is no cultural imperative that says we are going to make schlock."

*By late 1981, the number was up to one thousand companies.

Under pressure, American executives are rapidly changing their attitudes. One sign is the upgrading of quality-control officials. They have had little corporate heft, often answering to their natural enemies, the heads of manufacturing, who tend to be more interested in cost and quantity than quality. Chrysler has elevated its vice-president for quality so that he reports to the vice-chairman instead of the executive vice-president for manufacturing, and GM created the new post of vice-president in charge of quality and reliability. Ernst Beuler, Volkswagen of America's vice-president for quality assurance, brought from Germany an idea novel to Americans: if he is dissatisfied with the product, he can shut down the Rabbit line in Pennsylvania. Once close to blasphemy, this idea is spreading. At a critical point, when its new Escorts and Lynxes were beginning to come off the lines in 1980, Ford found a minor vibration in the transmission's torque converter. While engineers prepared a damping device, Ford shut down production of automatic-transmission models for three weeks.

Like all beginners, American management has more enthusiasm for excellence than skills to attain it. Hundreds of thousands of Japanese have been trained in quality assurance, but only a handful of U.S. colleges offer a degree in the subject. Consultants give seminars for management, and the American Society for Quality Control puts five to six hundred managers through its three- to five-day courses every year. But Juran thinks nothing short of "massive training" will do the job.

The experts like to say "quality is free." Anyone familiar with aerospace knows that isn't always so: exotic materials, redundancies, and 100 percent testing of components gobble up the taxpayer's money. Still, good quality that consumers can afford may actually reduce costs and almost always increases productivity. The earlier you detect a defect, the more you can save. Richard W. Anderson, gereral manager of the computer systems division of Hewlett-Packard, describes the damage a faulty two-cent resistor can do. If you catch the resistor before it is used and throw it away, you lose two cents. If you don't find it until it has been soldered into a computer component, it may cost ten dollars to repair the part. If you don't catch the component until it is in a computer user's hands, the repair will cost hundreds of dollars. Indeed, if a five-thousand-dollar computer has to be repaired in the field, the expense may exceed the manufacturing cost.

The more complicated products become, the more reliable components must be if costs are to be held down. Poor work affects expenses all along the line—in scrappage, repairs, larger inventories to provide a cushion against defective parts, higher warranty cost, and eventually lost reputation and sales. Although precise calculations are difficult, Juran estimates that the costs of poor quality can come to 5 to 15 percent of sales. In one factory Deming visited, the manager figured that 21 percent of the plant's capacity was tied up producing mistakes and correcting them.

LOOKING FOR PATIENT MONEY

Because the rewards of quality assurance are hard to measure, American businessmen have been slow to appreciate them. MBAs looking at estimates of discounted cash flow don't like improvements that take a long time to pay off—even though the payoff can ultimately be enormous, as the Japanese have proved. Stockholders are impatient too. Japanese and German companies are financed more by debt than equity, and the banks do not press them as hard as shareholders for inexorable quarterly earnings gains. If U.S. corporations are to become basically quality-oriented, they will need, perhaps more than anything else, patient money. They will need to rearrange the incentives that motivate managers.

It may take ten years to catch up with the Japanese, but evidence of progress is already visible. In March 1980 Hewlett-Packard announced the results of tests it ran on semiconductors. The failure rate of U.S. chips was five to six times that of the Japanese. The news apparently shocked the U.S. semiconductor industry into action. When Hewlett-Packard ran similar tests later in the year, the U.S. manufactures had cut their disadvantage to three to one.

Back in 1975 American color TVs failed three times as often as Japanese sets, according to one consumer survey. So the 1979 figure of 50 percent more failures represents a big improvement. After admitting frankly that the Japanese made better TV sets, Magnavox committed itself in 1981 to producing a set designed for quality from the ground up. The Phoenix, as it is to be called, will have 30 percent fewer parts than current sets. Promises a Magnavox executive: "It's going to be as good as anything the Japanese make, by design, by commitment, and by God."

December 29, 1980
Research associate: Patricia A. Langan

14

SERVICE WITHOUT A SNARL

JEREMY MAIN

Everybody has his favorite horror story—about being bumped in Puerto Rico and then hanging around the airport for a day to get another flight, or about the bank that turned a $60 check into a $600 check and took a month to correct the mistake, or about the plumber who connected the toilet to the steam heating. Inevitably the stories reinforce the old saw that you can't get good service anymore.

W. Edwards Deming, the American statistician who taught quality assurance to the Japanese, agrees. "Even the sloppiest manufacturing concerns have some idea of how they're doing," he says. "But in the service industries they don't even know when they're in trouble, and if they did they wouldn't know what to do about it." In a newly published study, the American Enterprise Institute says of the utility and transportation industries: "Service is not what it used to be, nor will it long be as good as it is now."

Are all those aids and comforts that make life livable really deteriorating? The question seems to defy a rational answer, since people feel so passionately about their particular misadventures that they tend to forget good experiences. Certainly if you still expect humble attendants to wait on you deferentially you will think service has gone to pot. The bank manager no longer steps out to greet you by name.

However, the bank does have twenty-four-hour automatic tellers and electronic fund transfers and other striking improvements in service. The airlines have nationwide computerized reservation systems that are models of convenience and efficiency. Only a curmudgeon could argue that the American telephone system, with its direct dialing to Bangkok and its fiber-optic cables, doesn't provide service that's better than ever.

"You can't deliver service in the old way, but new systems can actually deliver it better," says Richard M. Kovacevich, a senior vice-president who is using electronic razzle-dazzle to try to capture customers for New York's Citibank. "People tell us: don't be polite—be efficient, fast, and knowledgeable." Considering that Americans now write about forty billion checks a year, ritual politeness over each transaction would seem a little ponderous. In any case, Citibank's clients would probably agree that on the whole its electronic employees are more amiable than its human employees.

Citibank is only one of many large service businesses that have set out to prove Deming dead wrong. Beginning belatedly in the 1970s, these companies found they could improve service in much the same systematic way that manufacturers can improve their products. There are differences, of course. You can't build up an inventory of services. If you can't produce the service the moment it is needed you are likely to lose the sale. But many of the principles of quality assurance apply as much to services as to manufacturing. The whole enterprise, from top management down, must be dedicated to quality. Statistical controls must be used to pinpoint weaknesses and then monitor the corrections. The results of such scrutiny shouldn't be used to punish employees, but to help those who need it and reward those who deserve it. Where possible, machines should replace people in routine tasks because they are less error-prone.

IT'S NOT THE BROADWAY LIMITED

These principles of quality won't bring back a trip on the *Île de France* or revive the old Broadway Limited. The Big Mac may not be the best hamburger around, but it represents quality in the sense that you can rely on getting the same Big Mac anywhere in the U.S., served quickly and cheaply in clean, cheerful surroundings.

Quality assurance can show results just as striking in a service as in a factory. Error rates in processing checks in banks or doling out medicine in hospitals, for example, can be cut to a fraction of what they usually are. Like the average factory, the average office spends perhaps 20 percent of its effort making and correcting errors. The modest cost of quality controls is far less than the costs of mistakes.

Only large organizations, of course, have the resources to apply quality assurance systematically. Small organizations—and most of the forty-nine million nongovernment service workers in the U.S. are employed by small organizations— have a different type of advantage. A conscientious and competent boss can supervise the whole business and make sure it provides first-class service. But the incompetent can easily get into service businesses. It's possible to become a travel agent, for instance, with little capital, no bond or license, no training, and little knowledge of the places you're going to send your clients—maybe just the desire to get to see them yourself for free. Judging by the number of complaints they arouse, the home- and auto-repair businesses are among the worst of services. The high incidence of complaints about rip-off contractors and botched or unnecessary car repairs hasn't changed for years. It would be unrealistic to expect much quality control in these services other than the descretion the consumer applies to his choice of servicemen.

But poor performance is still widespread in large organizations as well as small. The postal service and the railroads rank down among home contractors in the public esteem. A traveler in Colorado recently found out how bad a big organization can be. She arrived at Denver's Stapleton Airport planning to catch a Trailways bus to the ski resort at Copper Mountain. She hadn't been able to reserve a seat because she couldn't get through to the Trailways 800 number. She waited for the next twenty-four hours, standing in lines, getting misinformation from unhelpful clerks, calling Trailways numbers that never answered, and generally being treated like an unwelcome visitor to medieval Bulgaria. Finally the Trailways people put her on a bus to Frisco, Colorado, where, they assured her, she could catch a shuttle to Copper Mountain. Naturally, at Frisco there was no shuttle. "We haven't had a shuttle since August," said the Trailways agent. "We've been trying to tell 'em that in Denver, but they never answer their phone."

DISPENSING WITH ERROR

In this case, poor quality control resulted in no more than a lot of aggravation, but even when the stakes are higher, quality control can be lax.

Hospitals are surprisingly unsuccessful in controlling the safety and efficacy of their treatment. Quality assurance falls through the gaps between the jealous jurisdictions of doctors, nurses, administrators, pharmacists, and technicians. When the Olympian doctor issues his orders he assumes they will be carried out. Often they are inadvertently amended as they pass through the different jurisdictions. Dr. Kenneth N. Barker, head of the department of pharmacy-care systems at Auburn University in Alabama, has been studying hospitals and nursing homes for years, noting what medication the patient gets and then going back to the doctor to find out what he prescribed. The error rate, Barker says, comes to about 10 percent. According to his findings, the errors are just as likely to be made with dangerous drugs as they are with milk of magnesia.

The Health Care Financing Administration, one of the many arms of the Department of Health and Human Resources, seems to have been impressed by Barker's evidence. The agency is drawing up new standards that nursing homes must meet to qualify for Medicare and Medicaid reimbursements. One proposed standard would have limited nursing homes to a medication error rate of 5 percent. Barker told the government that hardly a nursing home and not many hospitals could meet that standard. What regulation will emerge, if any, now that the Reagan Administration has taken over, isn't clear.

Proper quality control, says Barker, could bring the error rate down to less than 2 percent. Hospitals and nursing homes should adopt a method of distributing medicine called "unit-dose dispensing." Under this system, instead of sending up batches of medicine to the nursing stations, hospital pharmacies send up only each specific dose as needed, labeled and ready to administer. Many hospitals have adopted the unit system with good results. In the meantime, says Barker, patients in other hospitals can expect, on the average, one error per day in their medication. The fact that most patients survive this mistreatment attests either to the fortitude of the human organism or the irrelevance of a lot of medication. But the results are serious. "There's no question that people die of errors in medication," says Barker. "A lot of rich lawyers can testify to that."

Contrast the hospital or bus line with an organization that takes quality control seriously. The next time you are in an American Airlines lobby, see if a man with a stopwatch and clipboard is hanging around. He's there to see how long it takes you to get your ticket—the company standard says 85 percent of the passengers should not have to stand in line more than five minutes. When you land, you may find another such fellow checking to see how long it takes to get the bags off the plane.

American Airlines employees are held to dozens of standards—and checked constantly. Reservation phones must be answered within twenty seconds; 85 percent of the flights must take off within five minutes of departure time and land within fifteen minutes of arrival time. Cabins must have their proper supply of magazines. Performance summaries drawn up every month tell management how the airline is doing and where the problems lie. If a late arrival was caused by air controllers, that can't be helped. But an outbreak of dirty ashtrays may be traced to a particular cleanup crew. The manager responsible for the crew will hear about it. His pay and promotion depend on meeting standards. If he fails to meet them three months running without extenuating circumstances, he may be looking for a job.

Constant checking has helped make American Airlines the preferred domestic line in the 1980 Airline Passengers Association survey. (United Airlines ran a fairly close second and Eastern Air Lines, despite all those ads by chairman Frank Borman, topped a companion poll on airlines people try to avoid.) American has a good on-time record too. On major domestic routes monitored by the Civil Aeronautics Board, 92.7 percent of American flights were on time in November 1980, compared with only 54.7 percent of Pan American/National flights.

To keep track of the competition, American inspectors clock the performance of United and TWA. American tries to keep slightly but not too far ahead of them. "We can't afford to be a whole lot better," says American vice-president Jerry R. Jacob.

The price of fuel is what stops airlines from getting a whole lot better. The huge increases in fuel prices are literally squeezing the passenger. Typically, the number of seats abreast in economy class in a Boeing 747 has been increased from nine to ten. With this and some other changes, the average 747 packs in 410 instead of 360 passengers. Many aircraft will soon be equipped with slimmed-down seats that will reduce the front-to-back space occupied by each seat from thirty-four to thirty-two inches. No amount of efficiency or courtesy is going to make a cramped passenger feel good about the service. Seating is the focus of passenger complaints today.

It's hard to see how else the airlines can cope with inflation. Two decades ago, when they introduced the jets, the airlines achieved a huge improvement in productivity and, consequently, service. But they have nothing like the jets up their sleeves today. Inflation can be particularly hard on service companies.

The future for railroad passengers is grimmer. Without new technology, without enough capital, without good management, the railroads have little hope of rising above mediocrity. It's true that Amtrak has acquired new rolling stock and has edged closer to running its trains on time. Amtrak is even developing a passenger's bill of rights, which will include—improbable as it sounds—overnight shoeshines for passengers in sleepers. But the huge subsidies that have made Japanese and European passsenger services so good are unlikely here. Less than one-half of one percent of intercity travelers use the rails, and the Reagan Administration wants to cut Amtrak and Conrail subsidies.*

In contrast to the railroads, the American Telephone & Telegraph Company has everything that it takes to be a first-class service company—the best in technology, ample capital, and a well-practiced quality-control system. Every month, the Bell System and its twenty-three operating companies are rated in a widely distributed "green book" on the basis of 130 separate criteria. These include how long it takes to get a dial tone (it should not be more than three seconds 99 percent of the time), how well appointments to install phones are kept, the numbers of accidents involving employees and vehicles, and the operating margins at PhoneCenter stores.

The Bell System has had such internal measures of quality for decades. To these, the system added in the 1970s surveys of customers to find out how the service looked from the outside. Research firms engaged by Bell polled 4.2 million customers in 1980. The companies compete so closely for good ratings that there's little to separate the best from the worst. Indeed, that year the New York Telephone Company ranked last in courtesy one month and first in another. The ratings can be broken down within the companies to districts and even to individual switchboards and PhoneCenter stores. Managers down to the level of foreman are judged and paid at least in part on the basis of these ratings.

The other assets of the Bell System are perhaps even more impressive than the intensive monitoring. Every three weeks, year after year, in carefully scheduled cadence, one division after another of the Bell System dips into the capital market. Huge sums are raised for satellite communications, transmission circuits that can accommodate 11,000 calls on a single cable, and solid-state switchboards. It is as if all technology was converging to make phone service better. Computers are the heart of the improvements and have enabled the system to increase its service

*The Amtrak subsidy was actually cut by 26 percent and the Administration proposed to reallocate a $320 million subsidy for Conrail.

while cutting the number of operators almost in half. Better prices as well as better service result. Most calls cost less today in real dollars than they did thirty years ago. The computer makes a further contribution by automatically monitoring the quality of services. For instance, computers keep constant track of delays in dial tones.

In commercial banking, computers are not only enabling the banks to cope with all that paper but are changing the nature of the service the banks provide. The nation's largest bank holding company, New York's Citicorp, is spending hundreds of millions of dollars to conquer the consumer market with a strategy that relies more on machines than on face-to-face personal service. Citibank is concentrating its efforts on electronic banking—automated tellers that function day and night all week—and computerized monthly statements that are more readable and complete.

"Customers want to deal with people they know and that's not possible anymore," says Citibank's Richard Kovacevich. "But there is no reason why a terminal can't know more about you than a person can. If good computer records are available, there's no need to know you personally. Transactions as boring and mundane as collecting and dishing out money can be done by machines. We can get to politeness when we really need to deal personally with customers." Citibank customers who use the automatic tellers seem delighted. Those old-fashioned enough to prefer a little personal attention are not.

ATTRACTING THE DEADBEATS

Computerized quality controls helped Citibank recover from a spectacular disaster in 1977 and 1978, after the bank sent out mass mailings inviting twenty-six million people to accept bank cards. Five million responded, and Citibank was swamped with bad debts and other problems a hastily gathered staff couldn't cope with. The staff now operates smoothly from an automated phone room in Huntington, Long Island. When a customer phones with a problem or question, an operator can instantly display the last three months' activity in the customer's account on a screen. The operators used to have to rummage around looking for paper files, often taking so long the customer would hang up. When the customer tried again later, he would get another operator who, of course, would be unable to find the file the first operator had removed. Repeat calls—which imply that the customer was unable to get an answer on the first call—have been cut from 25 to 4.1 percent of the total. The operators work in teams of eight or nine each and are constantly monitored. Team-of-the-month awards go to those that perform best.

Citibank has yet to make a profit on the card venture. In fact, it lost about $100 million in 1980 because it was paying an average of 14 percent in annual interest for money to lend to card customers, while the New York usury law didn't let it charge more than 12 percent. The limit has since been lifted. In any case, Citibank seems willing to take big losses in this and other consumer ventures in order to grab a strong position in the growing market for automated banking.

Behind the scenes, banks are much like factories. Instead of processing drill bits or gaskets, they process bits of paper in vast numbers. Electronic funds transfers, which are faster and have an error rate at least a third less than that for paper transfers, are a growing but still small part of the process. Paper is still the basic medium of bank transactions. Years ago the banks industrialized their paper handling. They use machines like the IBM 3890 Document Processor, which zips through 70,000 checks an hour. But only in the 1970s did the banks begin to apply the kinds of quality controls that manufacturers use. When William J. Latzko came to New York's Irving Trust Company in 1970 as its first quality-control officer, he found the bank's management disturbed because the machines were rejecting so many checks—though nobody knew what the percentage was. Using standard statistical methods of quality control, Latzko found out how many checks were being rejected (it was 7½ percent) and why. It turned out that certain check manufacturers were doing a poor job of printing magnetic characters, and some of the encoding machines were unreliable. Once these and other problems had been corrected, the rejection rate dropped to less than 2 percent.

The insurance industry, another major processor of bits of paper, has been slower to adopt quality controls. However, when ITT took over the Hartford Insurance Group a decade ago, it started bringing in the kind of controls used in ITT's manufacturing operations. Frank Scanlon, director of quality, figures the company will be saving more than $7 million a year during the 1980s by reducing errors, particularly errors in rating insurance risks. The computers at the Hartford are educated not only to process transactions accurately but to spot mistakes made by their human operators.

FOREVER WHOLESOME

For all the improvements made possible by technology, the quality of service still often depends on the individual who delivers it. All too often he is underpaid, untrained, unmotivated, and half-educated. Even the mighty oil industry may be represented before the public by a grubby, indolent teenager who seems more interested in picking his nose than coming over to the pump to help.

A few of the biggest service companies manage to be pleasantly efficient, even when they have thousands of employees dealing with millions of customers. Visitors to Walt Disney World near Orlando, Florida, come home impressed with its cleanliness and the courtesy and competence of the staff. Disney World management works hard to make sure the 14,200 employees are, as a training film puts it, "people who fulfill an expectation of wholesomeness, always smiling, always warm, forever positive in their approach."

Even for dishwashers, employment at Disney World begins with three days of training and indoctrination in an on-site center known as Disney University. Disney doesn't "hire" people for a "job" but "casts" them in a "role" to look after the "guests" (never "customers"). If employees are to work in costume in the Magic Kingdom, they are told not to hesitate to go underground to the vast wardrobe for a fresh costume if the one they're wearing gets soiled during the shift.

They learn that everyone pitches in to make Disney World work. When the crowds get too big to handle—a record 93,000 visitors turned up on New Year's Day 1981—even the top managers and their secretaries leave their offices to work behind counters or in ticket booths. New employees begin to pick up the pride that goes with working for a first-class organization. They also learn the foibles. The late Walt Disney, who wore a toothbrush mustache all his adult life, insisted that male employees be clean-shaven, and they still are. Periodically, employees return to the classroom for more training and indoctrination—a process some of them call "fairy dusting." Thousands of employees try to get ahead every year by taking courses ranging from carpentry to college-level business management.

The secret of keeping Disney World so clean is never to let it get dirty. Disney himself set an example by picking up rubbish when he walked around. Employees on "potty patrol" have walkie-talkies so they can radio for help to keep the toilets clean on busy days. Every night the Magic Kingdom is scrubbed squeaky clean. Crews steam-clean the streets, polish every window, shine all the brass. The results, as Mary Poppins would say, are supercalifragilisticexpialidocious.

COOKING BY THE SAME BOOK

McDonald's may not be everyone's idea of the best place in town to dine, but at its level McDonald's provides a quality of service that is the envy of the industry. Whether you go to the McDonald's on Queens Boulevard in New York or the one in Elk Grove Village near Chicago's O'Hare airport, you know exactly what you will get. They all go by the same book. Cooks must turn, never flip, hamburgers one, never two, at a time. If they haven't been purchased, Big Macs must be discarded ten minutes after being cooked and French fries in seven minutes. Cashiers must make eye contact with and smile at every customer. Exact specifications alone aren't enough, however. The help, mostly eager youngsters, must also be motivated to perform a monotonous, low-paying job with sustained enthusiasm. Debbie Thompson, who started out at McDonald's as a cashier eight years ago and now, at 24, manages the company-owned store at Elk Grove Village, sometimes livens up the lunchtime rush hour by offering $5 bonuses to the cashiers who take in the most dollars and handle the most customers. She gives a plaque to the crew member of the month. Ira Meyer, a onetime stockbroker who owns the Queens store, gives every employee a daily rating on a scale of one to five. Like Debbie Thompson, he sometimes puts the store on "sixty-second service": any customer not served within sixty seconds of placing an order gets free French fries. "You've got to create excitement and instant recognition," says Meyer.

McDonald's, Citibank, and Disney World have certainly established new standards of modern service. At least they have found efficient means of dealing with masses of customers with a smile—perhaps a little glassy—at a reasonable cost. Their service bears the same relationship to classic service as a pair of jeans bears to a Chanel suit. Classic service can still be found, of course. The Hotel Algonquin in New York looks and acts much as it did in the days of the famous literary round table in the 1920s. Beneath the surface it has changed; it is equipped with smoke detectors, walkie-talkies, magnetic cards instead of room keys, and the best in modern beds and pillows. But the Algonquin doesn't flaunt these or other qualities. It relies still on the cozy atmosphere of its paneled rooms and the uncondescending excellence of its staff. "We don't think in terms of efficiency," says managing director Andrew A. Anspach. "We think in terms of maintaining a civilized, literate, comfortable, inn-like hotel. We are well aware we spend more than we might, but we are family-run and don't have to worry about the budget."

The two hundred employees know each other by name. "They're our secret weapon," says Anspach. "They are not so much servants as well-adjusted people who enjoy making others happy." Since the Algonquin has no mortgage and there are no stockholders pressing for maximum profits, the hotel isn't even too expensive. Singles, admittedly a bit cramped, cost $68 to $76 a night. Classic service usually does come at a high price. But there will always be a market for it, perhaps even a growing market. There seems to be a trend toward building smaller, classier hotels.

Sometimes a chain of hotels may approach the level of service offered by an Algonquin. Westin Hotels (known until January 1981 as Western International Hotels) operates fifty-five hotels in the U.S. and abroad that offer unusual amenities. Room service runs twenty-four hours a day. The chain guarantees a free room in another hotel to any guest with a reservation who is turned away. These are company-wide standards. Perhaps more important, each hotel strives to retain its individuality. At the St. Francis in San Francisco an employee washes all the coins in the tills overnight so the guests will not have to face soiled silver first thing in the morning. The Renaissance Plaza in Detroit is equipped with a handsome, and free, health club and pool.

Whatever happens to the quality of service, classic or canned, the customer can be relied on to keep complaining. "There's an almost insatiable appetite for service," says Citibank's Kovacevich. "Nobody ever says you have given me all the service I want. People always want better." There's no limit to impatience either. However much a service is speeded up, customers want it done faster. The children of Americans who once thought it marvelous to cross the continent by train in three days will be so impatient at the end of a 5¾ hour flight from New York to Los Angeles that they will jump out of their seats before the plane stops rolling and grab the bags they carried aboard because checking them might have cost another ten or fifteen minutes.

March 23, 1981
Research associate: Patricia A. Langan

15

SHARPENING THE COMPETITIVE EDGE

WILLIAM BOWEN

The U.S. in the late twentieth century is not much of a country for consensus, but even so, a public-opinion survey in 1980 found 90 percent agreement that the American economy was "seriously off on the wrong track." On that proposition, at least, the nation is united.

This sense that something is seriously wrong with the economy comes not only from the workaday pains of inflation and near-zero growth in real wages, but also from the perception that the U.S. is slipping: that its ability to compete effectively with other countries is seriously dwindling. In that same survey, conducted by Garth Associates for the New York Stock Exchange, more than 60 percent agreed that the U.S. was "losing ground compared to most other countries."

In more and more product lines, manufacturers in other countries seem to be able to offer combinations of price, quality, and service that win the preference of customers, not only in markets abroad but in the U.S. too. The devastation of the American automobile industry is only the most conspicuous symptom.

A strong but transient upswing in U.S. exports recently fuzzed the picture of waning competitiveness. This surge in exports reflected the depreciation of the dollar in 1977–78 and large capital-equipment purchases by some of our major trading partners. The longer-term pattern, though, is painfully clear: erosion of markets for American manufactures, at home as well as abroad. A recent analysis by trade specialists at the Department of Labor detected ominous signs of waning competitiveness in some product lines where the U.S. has been a strong exporter—power-generating machinery, pharmaceuticals, computers, even aircraft. Deterioration in market shares abroad, the study warned, could be a leading indicator of tougher competition from imports at home.

The task of restoring America's competitive edge will have to be a national undertaking, with roles for business, labor, and the public, but government policy has to provide the framework. A central objective of policy has to be expansion of business investment in plant and equipment. American industry has been underinvesting by the standards of the league it has to compete in.

BEHIND EVEN BRITAIN

During the 1970s our expenditures for plant and equipment ran to a little more than 10 percent of the Gross National Product, which put us last among the major industrial nations, behind even Britain. And that investment was even punier than it seemed. A significant portion went to meet requirements imposed by federal regulations. More important, the decade brought extraordinarily rapid growth in the labor force, and one result was slow growth in the capital-labor ratio, the amount of capital stock per worker.

The 1970s also, of course, brought huge rises in energy prices, and that too hurt the competitiveness of American industry. Even in the days of cheap energy, Japanese and European managements paid more attention to holding down energy costs than did their American counterparts. And since 1974 Japanese and European companies, with their higher levels of capital investment, have gone further than American companies in adapting to high energy prices.

Put all these considerations together, and it is plain that the U.S. has to start investing a larger share of its output just to keep from losing more ground. A reasonable objective would be to lift business outlays for plant and equipment to 12 percent of GNP within the next five years. That would put us in the European class, not far behind what West Germany has been averaging over the past decade.

Expanding productive investment by 2 percent of GNP would involve an enormous shift of resources—2 percent of 1980 GNP comes to well over $50 billion. Such a shift would make possible faster replacement or modernization of obsolete facilities, faster acquisition of advanced productive technology, and faster adaptation to high energy prices. In short, it would improve the nation's prospects of being able to compete in the 1980s and beyond.

Policies designed to lift the investment share of GNP would run into a tangle of impediments. First of all, there's tough social arithmetic to cope with: If investment is to absorb a larger share of GNP, the other claimants—consumers and government—must accept a smaller share. The policy package has to take account of that. Without measures to restrain other claims on the nation's output, an effort to increase the investment share would aggravate inflation by overheating demand. Investment increases demand for goods and services before it adds to supply. The new plant must be built, new equipment acquired and installed, and new workers trained and paid before any marketable output begins to emerge.

Another impediment is inflation itself, which retards investment in new plant and equipment in several ways. The high interest rates associated with high rates of inflation tend to discourage long-term commitments. Also, inflation compounds uncertainty about future costs and returns, so it tends to shift investment into safer-looking short-run undertakings, such as marginal changes in products or acquisition of existing assets, rather than into building up the basic capital stock. Accordingly, measures to encourage increased investment must be accompanied by a credible program to bring the rate of inflation down.

Subduing inflation requires government action to hold down on total demand for goods and services in the economy. As things are usually handled, unfortunately, such restraint on demand also

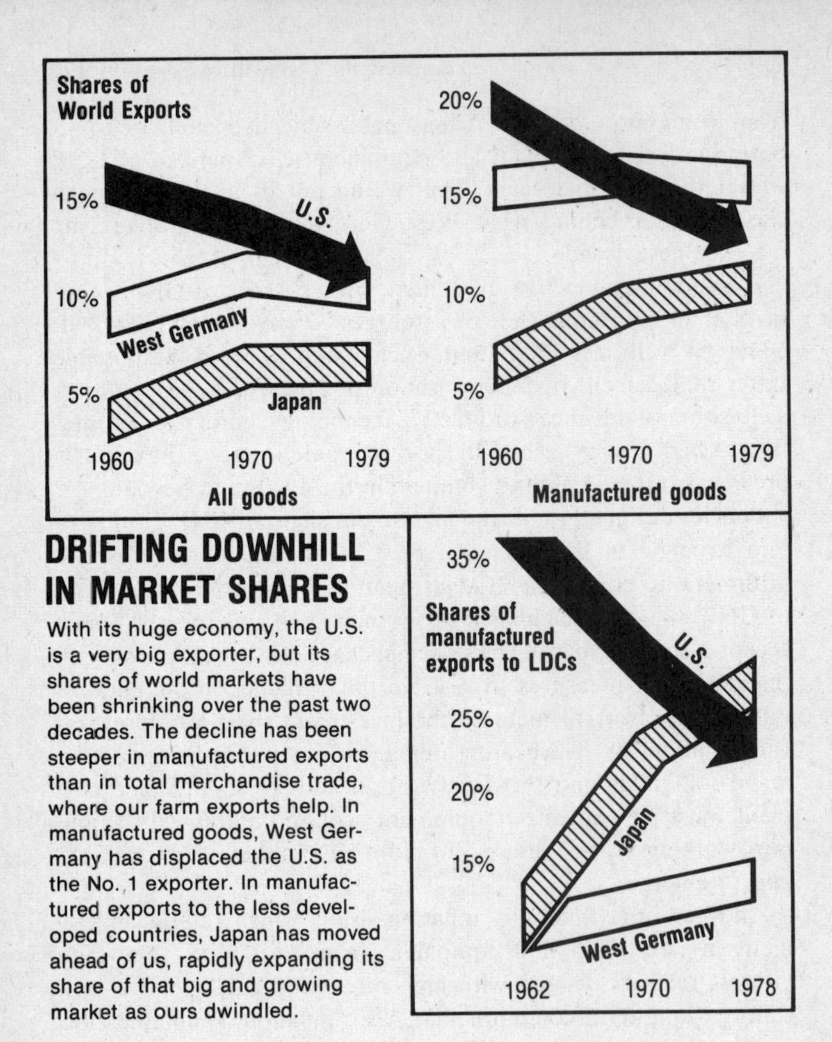

DRIFTING DOWNHILL IN MARKET SHARES

With its huge economy, the U.S. is a very big exporter, but its shares of world markets have been shrinking over the past two decades. The decline has been steeper in manufactured exports than in total merchandise trade, where our farm exports help. In manufactured goods, West Germany has displaced the U.S. as the No. 1 exporter. In manufactured exports to the less developed countries, Japan has moved ahead of us, rapidly expanding its share of that big and growing market as ours dwindled.

depresses supply. In a modern welfare state, government spending goes right on rising as demand slows down, so the burden of restraint falls entirely upon the private sector, and it falls especially hard on profits and investment. Demand restraint, then, not only holds down the current level of real output—that's unavoidable—but also holds down capital formation, thereby reducing future output.

SHIFTING THE SHARES

The escape ladder out of the pit is a coherent set of policies combining mild demand restraint with incentives to encourage capital investment—restraining present demand without dragging down future supply. *Fortune* proposes a policy framework consisting of four interacting components:

- Gradually reducing the growth rate of the money supply—moderately enough that investment won't be turned off by weakness in demand.
- Holding down hard on federal spending so that it grows less rapidly than GNP.
- Enacting tax incentives to encourage investment—incentives strong enough to persuade businesses to spend more on plant and equipment even though overall demand is only moderate.
- Altering tax incentives so as to induce Americans to save a larger proportion of their incomes.

A central objective is to shift the composition of GNP so that a larger share goes into productive investment and a correspondingly smaller share into consumption—government and consumer spending. The components of the program work together. The hold-down in public spending reduces the government's borrowing requirements, and this, together with the rise in the personal savings rate, makes it possible to finance additional business borrowing without strain on financial markets. The rise in the investment share of GNP increases demand for factory equipment, office machines, plant construction, and so forth, but this lift in investment outlays is balanced by moderation in spending for consumption.

In this way, a rise in capital investment can take place within an overall pattern of mild demand restraint. Much of the increased investment, perhaps most of it, will be of the "deepening" variety, primarily intended to improve efficiency rather than to expand capacity.

Further along the road, the higher level of investment would start paying off in increased supply and lower production costs, helping to bring down the underlying rate of inflation. Still further along, consumers would enjoy higher absolute standards of living than before, even though consumption would continue to account for a smaller share of the larger and faster-growing economic pie.

To put it mildly, this is not a formula for a quick victory over inflation. It is a formula for a slow victory—the only kind available to us. A slow victory is better than a series of illusory quick victories followed by retreats. The most important thing is to maintain a steady course, and not oscillate between too much restraint and too much stimulus—the woeful pattern we have become accustomed to.

At a glance, the proposed package may look much like the one shaping up in the new Administration, featuring tax cuts, reductions in federal spending, and moderate growth in the money supply. But there are important differences. As compared with the program proposed here, the emerging Administration program puts much more emphasis on broad income-tax cuts and less emphasis on measures specifically aimed at encouraging business investment and personal saving.

It remains to be seen, moreover, how serious the new Administration will be about fiscal discipline. Some Reagan advisers are passionately convinced that tax cuts would have powerful supply-side virtues even if unmatched by spending cuts. That is dubious. Unless largely offset by budget cuts, cuts in personal income taxes would not increase the funds available for investment. There would be a rise in consumer savings, of course, but the federal government's increased borrowing requirements, a result of the expanded budget deficit, would swallow up more savings than the tax cuts would generate.

PERVERSE MECHANICS AT WORK

The set of policies *Fortune* is proposing would be dauntingly difficult to carry out. The Fed's task would be easier with restraint in government spending, but monetary policy is an imprecise instrument at best. Getting the various elements of the package to work together would require more sophisticated economic management than is normally evident in Washington.

What's more, any serious attempt to rein in the growth of government spending would run into harsh political opposition. To a large extent, the political leverage of special-interest groups is what made nondefense spending grow so monstrously after the mid-1960s, and the same perverse mechanics will work to fend off any cutbacks. Every proposed cut, no matter how reasonable, brings howls from people who think they stand to lose something.

The Administration has got to decide just what programs it wants cut by just how much, then mobilize public opinion behind a vigorous effort to push the cuts through Congress. The effort might succeed. Since the huge distension of social-welfare spending set in under Lyndon Johnson, we have not had an Administration seriously dedicated to holding down federal spending. The 1980 election returns signaled strong public dissatisfaction with the way things had been going.*

Any program promising to wrestle down the inflation rate through some combination of fiscal and monetary restraint starts out with a low level of credibility. After the failures of past Administrations to achieve lasting gains against inflation, skepticism is certainly justified. But the particular approach advocated in this article has never been tried. Despite superficial similarities to some past programs, it is decisively different. Instead of pinching investment in the usual pattern, the program is designed to expand investment. And the restraint envisioned is sufficiently moderate to be sustained long enough to do some good. With the low-credibility problem, gains against inflation would probably be slow at first, but if the policies were held to resolutely, progress

*The 1981 Economic Recovery Tax Act that emerged from Congress after this was written soon aroused among investors the doubts expressed here. The stock market went into a severe slump in the fall of 1981. The Administration was soon looking for new ways to reduce spending and the budget deficit. And although consumers seemed to find good reasons to save more money in the tax revisions, business investment slumped along with the stock market. High interest rates, fear of a recession, and lack of confidence in the future continued to stifle capital spending.

would pick up after a while, as inflationary expectations cooled and help came in from the supply side.

The broad policy framework should be supplemented with other measures and programs to reduce business costs and improve productivity. Especially valuable would be reforms to cut the costs that regulation imposes on businesses. Almost every economist who scrutinizes regulation concludes that equivalent benefits could be obtained at much lower costs.

Where the problem is discharge of pollutants into air or water, the most cost-effective mode of regulation would be a system of pollution taxes (also called effluent charges or effluent fees). Tax is levied in proportion to pollution load, and it is left up to the polluter to work out the most cost-effective response. For each situation there would be a different optimum mix of pollution-control measures and tax payments. A major advantage of a pollution-tax system is that it sets no requirements regarding pollution-control technology, and therefore leaves open possibilities for innovation.

In occupational safety and consumer-product safety, among other matters, information could well be substituted for regulation. There is some doubt whether the Occupational Safety and Health Administration, in making an expensive nuisance of itself these past ten years, brought a worthwhile net gain in the safety of American workplaces—companies and unions, after all, had safety programs before there was an OSHA. The agency might accomplish at least as much for the safety of American workers if it confined itself to conducting research on safety methods, testing safety equipment, and disseminating the findings to company and union safety officers.

Proposals for enhancing productivity and competitiveness are being put forward in great abundance these days. If the U.S. economy is ailing, that is not for lack of prescriptions. A paper by John W. Kendrick, professor of economics at George Washington University, sets forth ninety-nine policy recommendations for promoting productivity growth. A search done for the Joint Economic Committee of Congress found no fewer than 514 distinguishable recommendations bearing on technological innovation.

A REMARKABLE DIVERSITY

Improvement in productivity performance is a bedrock requirement for enhancing America's ability to compete. With a few exceptions, notably autos and steel, our disadvantages in competition with Japan and Western Europe have little to do with high American wages. In fact, real wages in the U.S. have pretty much stagnated since 1973, while employment costs in Japan and Western Europe have risen substantially. The trouble is that American productivity growth has also stagnated.

The remarkable diversity in productivity among American companies suggests possibilities galore for improvement. Even companies turning out similar products show striking differences in productivity. When plants in a given industry are ranked according to output per worker, average productivity in the top quarter may well run three times as high as the average in the bottom quarter.

In part, these differences reflect the quantity and quality of capital per worker in the various companies. To some degree, however, the advantages of the high-productivity companies come from better industrial practices—more efficient work flow, tighter handling of inventories, less waste of materials. Better practices adopted in one company tend to diffuse through an industry, but usually the process extends over a considerable span of years. More rapid diffusion could bring catch-up gains in productivity and cost reduction.

J. Herbert Hollomon, head of the Center for Policy Alternatives at M.I.T., proposes a way to speed up the process: create an industrial counterpart of the Agricultural Extension Service, which has made a large contribution to the immense gains in American agricultural productivity over the decades. The Industrial Extension Service, as he conceives of it, would operate in a similar way, with practical-minded agents who work within a limited geographical area and get to know their territory. Along with good industrial practices, they could also disseminate up-to-date knowledge about quality-control methods, energy saving, and ways to improve employee motivation.

Our ability to compete in years ahead will largely depend on the pace of innovation in the U.S.—innovation that leads to new products or major improvements in old ones, or to distinctly better ways of doing things. Government can best encourage it by creating a favorable economic context, with less rapid inflation, less erratic interest rates, and less burdensome regulation. In the words of Hollomon's colleague James Utterback, who has made industrial innovation his special field, "A generally healthy climate for business is the most favorable climate for innovation."

More specific incentives to get American companies to devote more resources to research and development are also called for. R&D expenditures in the U.S. peaked in the 1960s at about 3 percent of GNP and have since declined as a share of GNP, partly because of the fall-off in government-funded military and space R&D. Over the same span, Japan and West Germany greatly increased R&D spending, and they have now about caught up with the U.S. at around 2 percent of GNP. Since military R&D bulks larger with us than with the Germans and Japanese, they may be putting proportionately more resources into commercially oriented R&D than we are.

To encourage privately funded R&D, some authorities on productivity recommend a special tax credit for additional R&D spending. Japanese companies get a tax credit equal to 20 percent of any year-to-year increase in R&D expenditures.*

THE FREE-RIDER PROBLEM

Market forces can't do everything, and one thing they don't do is bring about enough R&D in which the results consist of how-to knowledge of techniques rather than products that can be sold or processes that can be licensed. Such is the case with what is known as generic technology—technology that cuts across several industrial sectors. Albert A. Foer, an official in the Federal

*In 1981, Congress voted a 25 percent tax credit for certain corporate research expenditures that exceed the average of the previous three years. Congress also allowed corporations to depreciate capital equipment used in R & D in three years rather than ten.

Trade Commission's Bureau of Competition, explains that "certain types of generic research will not be undertaken because of the so-called free-rider problem: no one firm stands to benefit enough to undertake the expenditure, hence something the society could well find valuable may be left undone." Accordingly, some major industrial nations have a publicly supported program of generic-technology research. A notable example is the German research on numerical control, which has helped make West Germany the world's leading exporter of machine tools. Until 1980 the closest American counterpart was a Defense Department program on manufacturing technology. Then the Department of Commerce

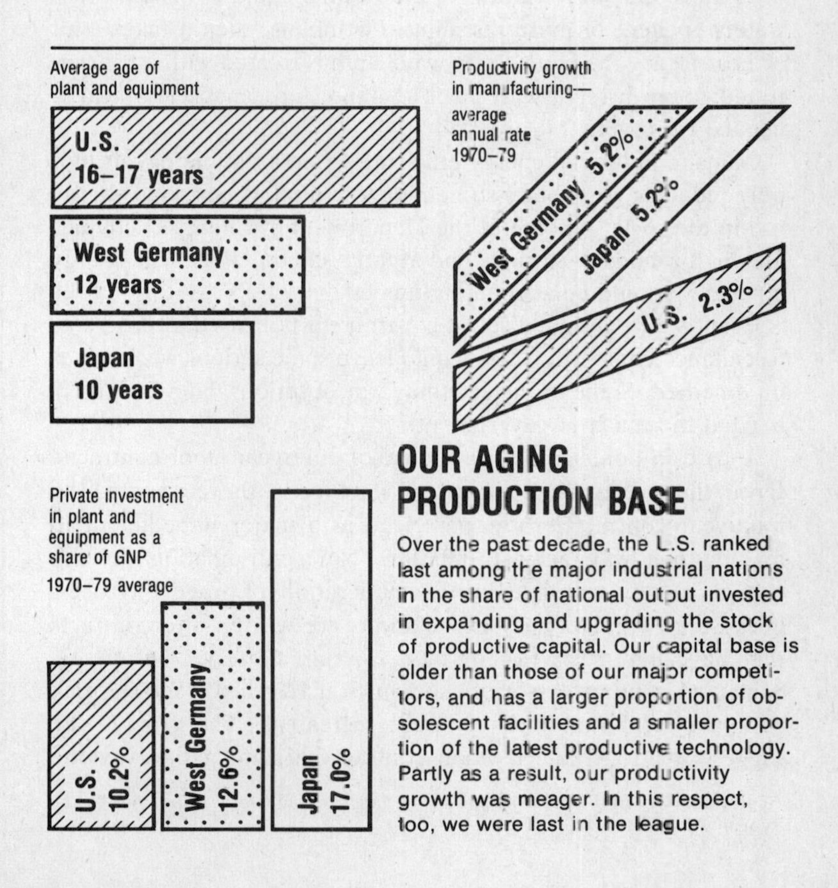

got off to a late but promising start with a program called COGENT—for Cooperative Generic Technology.*

The revolution in oil prices has made energy costs an important element of competitiveness. In response, the Department of Energy is running a broad program of R&D in industrial energy efficiency. Most of the projects are joint efforts, with private businesses doing the actual R&D and the Department of Energy providing guidance and part of the money.

The large payoffs come not from marginal improvements in old ways of doing things, but from leaps to new ways. An example of a leap comes from a project to reduce energy use in the finishing of textiles. Old way: cloth is treated with a liquid to make it, say, water-repellent or more resistant to wrinkling; then it takes a lot of heat to dry the cloth. New way: cloth is treated with a concentrated foam instead; with much less moisture, much less heat is needed for drying.

Gains in industrial energy efficiency bring a double payoff, not only reducing costs but also helping to hold down on total energy use in the society. Some of the Department of Energy's activities are in for budget cutting, and rightly so, but R&D on energy efficiency should be adequately funded.

Even a well-designed set of government policies and programs to enhance competitiveness would leave some serious weaknesses unremedied. Some of the economy's most serious flaws lie largely beyond the reach of government.

A case in point is the prevalence of multiyear labor contracts. Once thought to be a source of stability in the economy, the multiyear contract is now perceived as a major impediment to making headway against inflation. New anti-inflation policies cannot readily translate into a moderation of wage increases, because past inflationary expectations are written into contracts that have a year or two or more to run. Observes Jeffrey D. Sachs, a young professor of economics at Harvard: "We're stuck with a system that locks in wage increases." The remedy, he suggests, is to conduct union contract bargaining the way the

*Unfortunately, COGENT was one of the first programs to be wiped out by the Reagan Administration.

Germans and Japanese do: all contracts running for just one year, and the main ones coming up for renegotiation at about the same time.

It's a good idea, but not one that Congress could readily put into effect by passing a law, or the President by issuing an executive order. We are not likely to see much progress along this promising path unless enlightened corporate and union leadership takes up the cause.

NEGLECTING THE FUTURE

A number of observers have begun pointing to managerial failings as a major cause of the decline in competitiveness. One serious charge against American managers is that many of them do a poor job of enlisting employees on the side of improving productivity. Evidence suggests that the untapped potential may be quite substantial. In a survey of workers' attitudes, conducted by the Gallup Organization for the U.S. Chamber of Commerce, people were asked: "How much time have you spent thinking about changes that could be made in your company that would improve its performance?" Remarkably enough, 44 percent said "a lot of time" and another 18 percent said "some time"—results that would surprise many managers. It is a pretty safe bet that many of those workers had never been made to feel that their ideas on improving performance would be welcomed and appreciated, let alone rewarded.

Managers are often heard to complain that the quality of American workers is not what it used to be, but European managers say the same thing about *their* employees. And there is evidence that good management can improve the attitudes of employees—not by exhorting them, but by treating them as collaborators in the task of improving performance.

Elmer B. Staats, whose term as Comptroller General of the U.S. is about to expire, tells of a study of twenty similar coal mines in Wyoming. Production in tons per worker-day ranged from 58 to 242, though the companies were all mining the same kind of coal with much the same kinds of equipment under the same government regulations. "The main difference," Staats reports, "was how company management worked with its employees. The most productive firm provided its employees with the greatest amount of individual responsibility and involvement in decision-making."

Another charge levied against American managers is that they put too much emphasis on short-term results, and in doing so neglect the future. Looking for the roots of this bias, some critics point to the business schools, with their stress on current balance-sheet results as a measure of managerial effectiveness. Others point to the baneful influence of portfolio managers, under pressure to show short-term profits in order to attract and retain clients. The power of portfolio managers to affect the price of a company's stock, in turn, puts pressure on corporate managers to show short-term profits. Some of the critics also argue that basing executive bonuses on annual profits reinforces the short-term bias.

Since investment in the future detracts from the current balance sheet, the fixation on the short run works against the kinds of things that make for enhanced growth and competitiveness in the long run: building new plants, investing in R&D to create new products, patiently developing new export markets over a span of years, holding products back until they are thoroughly refined and tested. Over time, the consequences accumulate into serious competitive weakness.

AN EMPTY-BOTTLE WORD

Government cannot exert much direct influence on the perspectives of American managers, but something else could make a significant difference in a short span: the emergence of a consensus that strengthening America's international competitiveness is a compellingly important national task. Such a consensus could influence the attitudes not only of business executives but also of labor leaders, employees, intellectuals, even legislators and bureaucrats. The U.S. does not now have a widely shared and strongly felt sense of national purpose, and that is a serious handicap—just as Japan's sense of purpose has been a vital source of that nation's economic vigor.

Concern about America's competitive situation is much in evidence, but it remains free-floating, not attached to any particular program. The talk of "reindustrialization" reflects awareness of a problem rather than focus on a remedy. Sociologist Amitai Etzioni, who put the word into circulation a few years back, had something sensible in mind. He perceived that the U.S. had let its industrial base deteriorate relative to those of competing nations, and that rebuilding would require changes in attitudes and policies. But once in circulation the word lost specific content and became an empty bottle into which each user pours his own meaning—sometimes mostly fizz.

Having nothing much to say about substantive programs, those who wave the reindustrialization banner tend to stress the mechanics of consensus. Some put forward the notion that a grand tripartite commission representing business, labor, and government could be expected to come up with the answers. Some apparently think we should move toward becoming a sort of U.S.A., Inc., analogous to Japan, Inc.—as if the U.S. could somehow imitate the cultural and social characteristics that underlie the Japanese system.

The reindustrialization advocates are correct in their perception that consensus will be needed to give any national program of economic invigoration the sustained drive required to bring about major change. But a consensus that we need to do something is merely a shared uneasiness. To have any heft, a consensus has to be a consensus to do something in particular, follow a particular program, pursue a particular goal. The program does not emerge from the consensus—the consensus forms around the program.

The term reindustrialization blurs into another vague one—industrial policy—which is supposed to involve "picking winners" (something that markets can do much better than any government agency). As practiced by the Carter Administration, it had more to do with propping up losers—"lemon socialism," a phrase-maker dubbed it.

A notable case is the steel industry. To shelter it from foreign competition, the Carter Administration established a protective mechanism that in effect sets floor prices for imported steel. It also provided loan guarantees for several steel companies.

This can be a vicious trap for a nation to fall into. The British degraded their economy by channeling resources to noncompetitive industries in order to preserve jobs. By sheltering the American steel industry from competition, the government is ensuring that management will fail to make the strenuous efforts required to become competitive.

BETWEEN TOO CHANCY AND TOO COSTLY

Bad consequences seep outward into the economy. Protection for the steel industry raises costs for U.S. manufacturers who use steel, and that, in turn, makes *them* less competitive. Robert B. Reich, director of policy planning at the Federal Trade Commission, warns that further unpleasant effects lie ahead. "We may see the effect on American home-appliance manufacturers in the next few years," he says, "as Japanese appliances hit big." What the government should do about the steel industry is set a timetable—extending, say, five years into the future—for gradual withdrawal of protection from foreign competition. Concurrently, the government should adopt a set of policies designed to help companies in all industries, including steel, modernize productive facilities, hold down costs, and sharpen their competitive edge. Instead of getting into the chancy game of picking winners or the costly game of propping up losers, the government should try to create optimum conditions for the strengthening of competitiveness.

For many Americans, it is already clear that enhancing the nation's ability to compete in the world economy should be a major national objective. Businessmen make speeches about competitiveness, scholars write papers about it, Congress holds hearings on it, the executive branch issues reports on it. But the government also pursues numerous other aims, missing opportunities to improve competitiveness and doing things that impair it.

The U.S., as is often said, is a pluralistic society, and that pluralism extends to the matter of goals—let a hundred objectives bloom. In a competitive world, that won't do. Multiple goals on the same level get in each other's way. As they say at Texas Instruments, "More than two objectives is no objectives."

To guide policy and give it coherence, there has to be a very small number of central objectives—perhaps only one—that must be taken into account in the pursuit of all other objectives. We have been pursuing diverse aims with little or no regard for effects on competitiveness. What we should do, if we want to meet the challenge of the 1980s, is make enhancement of competitiveness a central objective and scrutinize all legislation and government actions from that perspective.

If the U.S. fails to develop a strong consensus on the need to compete, its relative position in the world in terms of influence and standard of living will decline. But if the U.S. resolutely decides to compete, it has some formidable advantages to bring to bear, including grand scale, better natural endowments than its major competitors, and a strong base of scientific training and research.

The picture was summed up well in a "consensus statement" issued by the Conference on U.S. Competitiveness, a gathering of businessmen, labor leaders, politicians, and academics at Harvard University in 1980. The statement warned that stagnating competitiveness "threatens both our economic health and our national security," and that "we will suffer continued decline unless we undertake very basic changes in our attitude and policies." All true. And then it added, right again: "An America ready to accept the challenge of staying competitive is on the road to recovery."

March 9, 1981
Research associate: Patricia Hough

BIBLIOGRAPHY

Following is a list of some of the books available on the subjects of productivity and quality:

Abernathy, William J. *The Productivity Dilemma: Roadblock to Innovation in the Automobile Industry*. Baltimore: The Johns Hopkins University Press, 1978.

Argyris, Chris. *Personality and Organization*. New York: Harper & Row, 1957.

Backman, Jules (ed.). *Labor, Technology, and Productivity*. New York: New York University, 1974.

Blake, Robert R., and Jane S. Mouton. *Productivity: The Human Side*. New York: AMACOM, division of American Management Associations, 1981.

Buehler, Vernon M., and V. Krishna Shetty, (eds.). *Productivity Improvement: Case Studies of Proven Practice*. New York: AMACOM, division of American Management Associations, 1981.

Cole, Robert E. *Japanese Blue Collar: The Changing Tradition*. Berkeley, Calif.: University of California Press, 1971.

———. *Work, Mobility, and Participation*. Berkeley, Calif.: University of California Press, 1979.

Crosby, Philip B. *Quality is Free: The Art of Making Quality Certain*. New York: McGraw-Hill, 1979.

Deming, W. Edwards. *Statistical Techniques for Improvement of Quality and Productivity*. Cambridge, Mass.: The M.I.T. Press, 1982.

Dyer, William. *Team Building: Issues and Alternatives.* Reading, Mass.: Addison-Wesley, 1977.

Fabricant, Solomon. *A Primer on Productivity.* New York: Random House, 1971.

Feigenbaum, A.V. *Total Quality Control Engineering and Management.* New York: McGraw-Hill, 1961.

Glaser, Edward M. *Productivity Gains Through Worklife Improvement.* New York: Harcourt Brace Jovanovich, 1976.

Gooding, Judson. *The Job Revolution.* New York: Walker and Co., 1972.

Greenberg, Paul D., and Edward M. Glaser. *Some Issues in Joint Union-Management Quality of Worklife Improvement Efforts.* Kalamazoo, Mich.: W.E. Upjohn Institute for Employment Research, 1980.

Greiner, John M., Harry P. Hatry, Margo P. Koss, Annie P. Millar, and Jane P. Woodward. *Productivity and Motivation, A Review of State and Local Government Initiatives.* Washington, D.C.: The Urban Institute Press, 1981.

Hayes, Frederick O'R. *Productivity in Local Government.* Lexington, Mass.: D.C. Heath and Co., 1977.

Heaton, Herbert. *Productivity in Service Oraganizations: Organizing for People.* New York: McGraw-Hill, 1977.

Hinrichs, John R. *Practical Management for Productivity.* New York: Van Nostrand Reinhold, 1978.

Holzer, Marc (ed.). *Productivity in Public Organizations.* Port Washington, New York: Kennikat Press, 1976.

Jenkins, David. *Job Power.* Garden City, N. Y.: Doubleday & Co., 1973.

Juran, Joseph M., and Frank M. Gryna, Jr. *Quality Planning and Analysis.* New York: McGraw-Hill, 1974.

Kendrick, John W. *Understanding Productivity: An Introduction to the Dynamics of Productivity Change.* Baltimore: The Johns Hopkins University Press, 1977.

Kendrick, John W., and E.S. Grossman. *Productivity in the United States: Trends and Cycles.* Baltimore: The Johns Hopkins University Press, 1980.

Kendrick, John W., and Beatrice N. Vaccara. *New Developments in Productivity Measurements and Analysis.* Chicago: University of Chicago Press, 1980.

Kerr, Clark, and Jerome M. Rosow (eds.). *Work In America: The Decade Ahead.* New York: Van Nostrand Reinhold, 1979.

Lefton, R.E., V.R. Buzzotta, and Manuel Sherberg. *Improving Productivity Through People Skills.* Cambridge, Mass. Ballinger Publishing Co., 1980.

Lesieur, Frederick G. (ed.). *The Scanlon Plan—A Frontier in Labor-Management Cooperation.* Cambridge. Mass.: The M.I.T. Press, 1958.

Likert, Rensis. *The Human Organization.* New York: McGraw-Hill, 1967.

———. *New Patterns of Management.* New York McGraw-Hill, 1961.

Mali, Paul. *Improving Productivity.* New York: John Wiley & Sons, 1978.

Maslow, Abraham H. *Motivation and Personality.* New York: Harper & Row, 1954.

Maynard, H. *Industrial Engineering Handbook.* New York: McGraw-Hill, 1971.

McGregor, Douglas. *The Human Side of Enterprise.* New York: McGraw-Hill, 1960.

Moore, Brian, and Timothy L. Ross. *The Scanlon Way to Improved Productivity: A Practical Guide.* New York: John Wiley & Sons, 1978.

Myers, M. Scott. *Every Employee a Manager.* New York McGraw-Hill, 1970.

O'Toole, James. *Making America Work* New York: Continuum, 1981.

Ouchi, William G. *Theory Z: How American Business Can Meet the Japanese Challenge.* Reading, Mass.: Addison-Wesley Publishing Co., 1981.

Pascale, Richard Tanner, and Anthony G. Athos *The Art of Japanese Management.* New York: Simon and Schuster, 1981.

Rosow, Jerome M. (ed.). *Productivity: Prospects for Growth.* New York; Van Nostrand Reinhold, 1981.

Ross, Joel E. *Managing Productivity.* Reston, Va.: Reston Publishing Co., 1977.

Schrank, Robert. *Ten Thousand Working Days.* Cambridge, Mass.: The M.I.T. Press, 1978.

Shaw, John C. *The Quality-Productivity Connection in Service-Sector Management.* New York: Van Nostrand Reinhold, 1978.

Siegel, Irving H. *Company Productivity: Measurement for Improvement.* Kalamazoo, Mich.: W.E. Upjohn Institute for Employment Research, 1980.

Stankard, Martin F. *Successful Management of Large Clerical Operations*. New York: McGraw-Hill, 1981.

Vogel, Ezra F. *Japan as Number One: Lessons for America*. New York: Harper & Row, 1979.

Walker, Charles R., and Robert H. Guest. *The Man on the Assembly Line*. New York: Arno Press, 1952 and 1979.

Washnis, George J. (ed.). *Productivity Improvement Handbook for State and Local Government*. New York: John Wiley & Sons, 1980.

Weber, Max. *Economy and Society*. G. Roth and C. Wittich, editors. New York: Bedminster Press, 1968.

Work In America. Report of a Special Task Force to the Secretary of Health, Education, and Welfare. Cambridge, Mass.: The M.I.T. Press, 1973.

APPENDIX

Following is a list of some of the research, academic, trade, professional, and governmental organizations that deal with productivity and quality:

American Center for Quality of Work Life · 3301 New Mexico Ave., N.W. · Suite 202 · Washington, D.C. 20016 · (202) 338-2933

American Institute of Industrial Engineers, Inc. · Committee on Productivity · 25 Technology Park/Atlanta · Norcross, Georgia 30092 · (404) 449-0460

American Productivity Center · 123 North Post Oak Lane · Houston, Texas 77024 · (713) 681-4020

American Productivity Management Association · 4711 Golf Road · Suite 412 · Skokie, Illinois 60076 · (312) 677-9141

American Society for Quality Control · 230 West Wells St. · Milwaukee, Wisconsin 53203 · (414) 272-8575

The Center for Effective Organizations · University of Southern California · Bridge Hall 200 · Los Angeles, California 90007 · (213) 743-8765

Center for Manufacturing Productivity and Technology Transfer · Room 5304 JEC · Rensselaer Polytechnic Institute · Troy, New York 12181 · (518) 270-6724

Center for Quality of Working Life · Institute of Industrial Relations · University of California · 405 Hilgard Ave. · Los Angeles, California 90024 · (213) 825-1095

Computer Aided Manufacturing International, Inc. · 611 Ryan Plaza Drive · Suite 1107 · Arlington, Texas 76011 · (817) 860-1654

Georgia Productivity Center · Georgia Institute of Technology · Engineering Experiment Station · Atlanta, Georgia 30332 · (404) 894-3830

Project on Technology, Work and Character · 1710 Connecticut Ave., N.W. · Washington, D.C. 20009 · (202) 462-3003

Industrial Development Division · Institute of Science and Technology · University of Michigan · 2200 Bonisteel Boulevard · Ann Arbor, Michigan 48109 · (313) 764-5260

International Association of Quality Circles · P.O. Box 30635 · Midwest City, Oklahoma 73140 · (405) 737-6450

International Council for the Quality of Working Life · Sussex Place, Regent's Park · London NW1 4SA · United Kingdom · 01-262-2918

Management and Behavioral Science Center · Wharton School · University of Pennsylvania · Vance Hall, 3733 Spruce St. · Philadelphia, Pennsylvania 19104 · (215) 243-5736

Manufacturing Productivity Center · Research Institute · Illinois Institute of Technology · 10 West 35th St. · Chicago, Illinois 60616 · (312) 567-4800

Maryland Center for Productivity and Quality of Working Life · College of Business and Management · Room 0121 Tydings Hall · University of Maryland · College Park, Maryland 20742 · (301) 454-6688

Michigan Quality of Work Life Council · 6560 Cass · Suite 315 · Detroit, Michigan 48202 · (313) 362-1611

National Center for Public Productivity · John Jay College of Criminal Justice · 445 West 59th St. · New York, New York 10019 · (212) 489-3975

Northeast Labor Management Center · 30 Church St. · Suite 301 · Belmont, Massachusetts 02178 · (617) 489-4002

Office of Personnel Management · Productivity Resource Center · P.O. Box 14108 · Washington, D.C. 20415 · (202) 632-6151

Oklahoma Productivity Institute · School of Industrial Engineering and Management · Oklahoma State University · Stillwater, Oklahoma 74078 · (405) 624-6055

Productivity Council of the Southwest · STF 124 · 5151 State University Drive · Los Angeles, California 90032 · (213) 224-2975

Productivity Institute · College of Business Administration · Arizona State University · Tempe, Arizona 85287 · (602) 965-7626

Productivity Research and Extension Program · North Carolina State University · P.O. Box 5511 · Raleigh, North Carolina 27650 · (919) 733-2370

Productivity Resource Center · Office of Productivity, Technology and Innovation · U.S. Department of Commerce · Washington, D.C. 20230 · (202) 377-1581

CIDMAC · Computer Integrated Design Manufacturing and Automation Center · School of Industrial Engineering · Purdue University · West Lafayette, Indiana 47907 · (317) 494-5400

Quality Circle Institute · 1425 Vista Way · Airport Industrial Park · P.O. Box Q · Red Bluff, California 96080 · (916) 527-6970

Quality of Working Life Program · Center for Human Resource Research · Ohio State University · 5701 North High St. · Worthington, Ohio 43085 · (614) 422-3390

Quality of Working Life Program · Institute of Labor and Industrial Relations · University of Illinois at Urbana-Champaign · 504 East Armory Ave. · Champaign, Illinois 61820 · (217) 333-0981

Quality of Work Life Center for Central Pennsylvania · Pennsylvania State University · Capitol Campus · Middletown, Pennsylvania 17057 · (717) 984-6050

Society of Manufacturing Engineers · One SME Drive · P.O. Box 930 · Dearborn, Michigan 48128 · (313) 271-1500

Texas Hospital Association · State Wide Hospital Productivity Center · P.O. Box 15587 · Austin, Texas 78761 · (512) 453-7204

The State Government Productivity Resource Center · The Council on State Governments · Iron Works Pike · P.O. Box 11910 · Lexington, Kentucky 40578 · (606) 252-2291

Utah State University Center for Productivity and Quality of Working Life · UMC 35 · Utah State University · Logan, Utah 84322 · (801) 750-2283

Work in America Institute, Inc. · 700 White Plains Road · Scarsdale, New York 10583 · (914) 472-9600

INDEX

Abernathy, William J., 95
aerospace industry, 30–31
affirmative action, 13
AFL-CIO, 60, 118, 127
age-sex composition of labor force, 6
Agricultural Extension Service, 185
Agriculture, U.S. Department of, 142, 154
aircraft industry, 19–20
Air Force, U.S., 22, 25
Airline Passengers Association survey, 169
Algonquin hotel (New York, N.Y.), 175–76
American Airlines, 168–69
American Center for the Quality of Work Life, 68, 126
American Cyanamid, 56
American Enterprise Institute, 164
American Productivity Center (Houston, Tex.), 1, 9, 16, 42
"American Renewal" (Bowen), x
American Society for Quality Control, 155, 161
American Telephone & Telegraph. *See* AT & T
Amtrak, 169–70, 170*n*
Anderson, Richard W., 161
Andrews, Marvin A., 148
Anheuser-Busch, 134, 139
Anspach, Andrew A., 175
Applicon, Inc., 29, 30, 32
Argyris, Chris, 65
Arthur D. Little. *See* Little, Arthur D.
Athos, Anthony G., 58
AT & T, 60, 124, 126, 170
Auburn University (Alabama), 167
automation, 22, 23, 71
automobile industry, ix, 57, 92–105, *94*; Japanese vs. U.S., 92–105, *94*

Bakery, Confectionary and Tobacco Workers International, 118–19
Bank of America (San Francisco, Calif.), 85–86
Barker, Kenneth N., 168
Battista, Anthony R., 145
Bell, Steve, 57
Bell System, 60, 121, 170
Beracha, Barry H., 132
Bernstein, Sanford C., & Co., 103
Beuler, Ernst, 161
BIF water-treatment equipment, 50
Big Mac, the, 165–66, 174
Bisanz, Charles, 68
Black & Decker, 53
Blough, Roger, 134
Blue Cross Center (St. Louis, Mo.), 138
Bluestone, Irving, 60, 120, 122
Boeing Aircraft, 52
Bonner, Boyd L., 102
Booz Allen & Hamilton, 45, 47, 148
Borman, Frank, 159
Boston University, 149

Bowen, William, x, 1–18, 177–93
Boyden Associates, 54, 55
Breckenridge, Don, 138
Breisch, Roger, 56
Brinner, Roger, 1
Brown, Andrew C., 92–105, *95–96*
Brown, Charles L., 60
Building and Construction Trades Council (St. Louis, Mo.), 130, 134
"buoyancy" factor, 10
Burck, Charles G., xi, 57–69, 92–105, 106–17, 118–27
Burger King, 33–34, 40–43
Burlington, Mass., 29
Butts, George, 100
Bylinsky, Gene, 19–32

CAD/CAM (Computer Aided Design/ Computer Aided Manufacturing), ix, 19–32, 51
Caldwell, Philip, 103
Calhoun, John A., 83, 84, 89
Callanan, Lawrence L., 133, 136
Callanan, Tom, 133
Calma, 30
Cambridge, Mass., 45
CAM-I (Computer Aided Manufacturing International Inc.), 25, 31
Canada, vii, 141–42
capital, fixed investment rate of, 7, 177–182
capital-labor ratio, 7, 185
Carnegie-Mellon, 54
Carter, President James Earl, 191, 192
Cenowa, Ronald A., 20
Center for Japanese Studies, University of Michigan, 92, 153
Center for Policy Alternatives, M.I.T., 185
Center for the Study of American Business, Washington University, 9
Chamber of Commerce, U.S., 67, 189
Chambers, David, 156
Chevrolet Gear & Axle (the Gear), 110–12
Chevrolet Motor Division, 102
Chrysler Australia, 99, 104
Chrysler Corporation, 93, 95, 99–100, 125, 128, 157, 159, 161
Cifelli, Anna, xi, 70–80, 81–91, 140–51
CIM (Computer Integrated Manufacturing), 24–26, 28
Cincinnati Milacron, Inc., 158, 159–60
Citibank (New York, N.Y.), 84–85, 165, 171–72, 175, 176
Citicorp (New York, N.Y.), 171
Civil Aeronautics Board, 169
Civil Service Reform Act, 146
Coates, William A., 72, 73–75, 77, 78, 79, 80
COGENT (Cooperative Generic Technology), 188, 188*n*
Cole, Robert E., 92, 98, 100, 153
Columbia University, 148; Business School, 54–56
Commerce, U.S. Department of, 187–88
Communications Workers of America (CWA), 60, 118, 124, 126, 127
Comptroller General, U.S., 189
Computer Aided Design/Computer Aided Manufacturing. *See* CAD/CAM
Computer Aided Manufacturing International Inc. *See* CAM-I
computer graphics, 20–21, 23, 28; three-dimensional modeling on, 28–29; wire-frame modeling on, 28–29
Computer Integrated Manufacturing. *See* CIM
Computervision, 30, 32
Conference on U.S. Competitiveness, 193
Congress, U.S., 9, 183, 192
Conrail, 170, 170*n*
construction industry (St. Louis, Mo.), 128–39
construction industry, labor disputes in, x
Consumer Reports, 154, 157*n*
Copper Mountain, Colo., 166–67
Corning Glass Works, 34–37, 43, 52, 53
Corning Pressware, 35–36
cost opportunities, 36
Council of Economic Advisers, U.S., 9*n*, 11
Council of Construction Employees, 134
Creason, Ted, 57, 58
Crompton Company, 34, 38–40, 43
Cummings, Thomas G., 113, 115
cyclical influences on productivity, 4–5

Dana Corporation, 66, 108, 112–13
Danforth, Douglas D., 71, 72
Data Resources, Inc., 1
Datsun, 154
Dayton, Ohio, 112, 125–26
Dayton (Ohio) Press, 126
Dearborn, Mich., 101

Deere & Company, 52, 53, 152
Defense, U.S. Department of, 187
Demaree, Allan T., xi
Deming, W. Edwards, ix, 155, 156, 160, 162, 164, 165
demographic influences on productivity, 5–6
Denison, Edward F., 3, 17
Denver, Colo., 166
Des Moines, Iowa, 33
Detroit. *See* automobile industry
Disney Productions. *See* Walt Disney World
Disney University, 173
Disney World. *See* Walt Disney World
Donahue, Thomas R., 60, 118
Dreyfus Corporation, 11
Drive-through Task Force (Burger King), 41
Durango, Colo., 72
Dur-Cor, 36
Dworkin, Peter, 1–18

Eastlake, Ohio, 53
Eastman Kodak, 60
Eastwood, Margaret A., 26
Economic Recovery Tax Act, 183*n*
economy, growth rate of, 9–10. *See also* GNP
Edison, Thomas Alva, 34
educational level of labor force, 6
Eisenberg, David, 103, 104
Elk Grove Village, Ill., 174, 175
Elliot, Jay R., 89–90
"employee-involvement" program, 103, 106–107
Energy, U.S. Department of, 188
Engelberger, Joseph F., 32
England. *See* Great Britain
Ephlin, Donald F., 121–22
Etzioni, Amitai, 67
Evans, David C., 23
Evans & Sutherland Computer Corporation, 23
executives in manufacturing, 55

factory management, ix, 44–56
Fallon, Walter A., 60
Federal Trade Commission, 153, 192; Bureau of Competition, 186–87
Feinstein, Dianne, 150, 151
First National Bank Building (St. Louis, Mo.), 138
Fiscal Advisory Committee (San Francisco, Calif.), 150–51
Fisher Body. *See* General Motors, Fisher Body division
Fleischer, Alfred J., 129, 130, 134, 135, 137, 138
Foer, Albert A., 186
Foley, Harold, 136
Ford, President Gerald R., 11
Fordham University, 73
Ford Motor Company, 93, 101, 103, 122, 123, 152, 159, 161
Fortune magazine, viii, ix, x, xi, 2, *4*, 7, 10, 33–34, 181, 182
Foster, John Stuart, Jr., 53
Fox, Karen M., 90
Franklin Park, Ill., 158
Fraser, Douglas, 122
Frigidaire, 126
Frisco, Colo., 167
Future of Business Regulation, The (Weiderbaum), 9

Galbraith, John Kenneth, 45
Gallup Poll, the, 67, 189
Garth Associates, 177
Gateway Center (Pittsburgh, Pa.), 70
Gateway Ammunition Plant (St. Louis, Mo.), 132
GE. *See* General Electric
Geier, James A. D., 158–60
General Accounting Office, 142, 144
General Electric (GE), 27–28, 30, 31, 152
General Foods, 107
General Motors (GM), x, 27, 29, 60, 66, 68, 93, *94*, 101–103, 108, 110–12, 116, 120, 121,123,125,156–57, 159, 161; Fisher Body division, 20, 27; Inland division, 125–26. *See also* Guide division
General Signal, 49–50
Genesco, 156
George Washington University, 2, 184
Germany, *62–63*, 161. *See also* West Germany
Gilbreth, Frank B., 62–63
GM. *See* General Motors
GNP (Gross National Product), 7–8, 10, 14, 178, 179, 181–82
Golden Gate Park (San Francisco, Calif.), 150
Good Morning America (ABC), 57

government productivity, x, 140–51
Grand Rapids, Mich., 76, 77, 79
Granite City Steel Company, 119
Grayson, C. Jackson, Jr., 1, 42
Great Britain, vii, *62–63*
Greenspan, Alan, 2–3, 7, 10
Grierson, Donald K., 31
Gross National Product. *See* GNP
group technology, 24
Grove, Andrew S., 83, 91
Guest, Robert J., 64, 67
Guide division (General Motors), 102

Hamilton, David J., 84
Harahan, William J., 100–101
Hardy, Jerome, 12
Harbour, James E., 95, 96, 97, 104
Harris, Louis, 67
Harris Poll, the, 142
Hartford Insurance Group, 173
Harvard Business Review, 45
Harvard University, 11, 14, 188, 193; Business School, 45, 46, 53, 54, 65, 95, 111
Hawthorne experiments, 64
Health Care Financing Administration, U.S., 167
Health and Human Resources, U.S. Department of, 167
Herman Miller, Inc., 108, 112
Herzberg, Frederick, 64
Hewitt, William A., 52
Hewlett-Packard, x, 44, 52, 61, 110, 161, 162
Hill, Charles E., 149
Hinsdale, Ill., 27
Hoglund, William E., 57, 58
Hogue, Thomas L., 90
Hollomon, J. Herbert, 185, 186
Honda, 154
Honeywell, 32, 46–47
Hootnick, Laurence R., 83, 84, 91
Horner, William, 122, 123
Hough, Patricia, 177–93
House Armed Services Committee, 145
Housing and Urban Development, U.S. Department of, 148
Houston, Tex., 43
Howitt, Thomas, Jr., 37
Hunn, Arthur A., 134
Hunt, James B., Jr., 141, 146–47
Huntington, Long Island, N.Y., 171
Hutchins, William "Red," 125, 126
Hyde Park, Mass., 78

IBM, 31–32, 61, 89; Data Processing Product Group, 51; 3890 Document Processor, 172
ICAM (Integrated Computer Aided Manufacturing), 25
Important Jurisdictional Disputes Board (Washington, D.C.), 137
Industry Products Company (Westinghouse), 73
inflation, 1–2, 179–82
Integrated Computer Aided Manufacturing. *See* ICAM.
Intel Corporation, ix, 81–91, *88–89*
International Association of Machinists, 111
International Association of Quality Circles, 109
International Brotherhood of Electrical Workers, 118
International Business Machines. *See* IBM
International Company (Westinghouse), 73
International Harvester, 25–26, 27, 152
intersectoral shifts, 8–9
Irving Trust Company (New York, N.Y.), 172
ITT, 173

Jacob, Jerry, R., 169
Japan, vii, ix, 32, 44, 51, *63*, 70, 71, 72, 73–74, 108, *180*, 185, 186, *187*, 191
Japanese quality standards, 153
Johnson, President Lyndon B., 133, 183
Johnson, S.C., & Son, 47–48
Joint Conference Board (Toledo, Ohio), 130
Joint Economic Committee, U.S. Congress, 1, 12, 140, 184
Jorgenson, Dale, 14
Juran, Joseph M., ix, 155–56, 161
"just-in-time" system, 97–98

Kanter, Rosabeth Moss, 68
Kehre, Howard H., 103
Kendrick, John, 2, 17, 184
Kennard, George F., 51
Kohnen, John, 132
Korn/Ferry, 54
Kovacevich, Richard M., 165, 171, 176
Kraft, Richard A., 159

Labor, U.S. Department of, 178
labor force, composition of, 6
Labor Statistics, U.S. Bureau of, 3, 4, 8–9, 142, 144
Landen, Delmar L. "Dutch," 120
Lang, Michael, 9
Langan, Patricia A., 152–63, 164–76
Langhorst, Ollie, 138
Latkowski, Doug, 111–12
Latzko, William J., 172
Lawrence, Kans., 114
Lawrence, Paul, 111
Leesburg, Ala., 39
Likert, Rensis, 65
Little, Arthur D., 45, 96, 160
Livermore Laboratory, 53
Lloyd, Bette J., 87
Lockheed Aircraft, x, 107
Lockheed Georgia Company, 31
Lombardi, Vince, 73
Lord, William, 38, 39
Los Angeles, Calif., 109, 154
Lubar, Robert, 44–56

MacAvoy, Thomas C., 35
Maccoby, Michael, 124, 126
Magnavox, 163
Main, Jeremy, xi, 70–80, 81–91, 140–151, 152, 164–76
Mair, Alex C., 101, 102, 156–57
Malkiel, Burton, 1
Malo, Alfred, 50
Management and Budget, U.S. Office of, 3, 146
Man on the Assembly Line, The (Walker and Guest), 64
Manoogian, John A., 152
Mantia, Richard, 129–30, 132, 134, 136, 137, 138, 139
"Manufacturing in the Eighties" (conference), 53
manufacturing executives, 44–56
"Manufacturing—Missing Link in Corporate Strategy" (Skinner), 45
manufacturing mission, the, 47
Marines, U.S., 71
Mark, Jerome A., 8
Maslow, Abraham, 64–65
Massachusetts Institute of Technology. *See* M.I.T.
"matrix" management, 71
Matsushita Electric Industrial Company, 158–59
Matthews, Ronald, 36
Mazda, 154
McBride, Lloyd, 119–20, 124
McCarthy Bros. Construction Company, 129
McCarthy, Timothy R., 129
McDonald, F. James, 60
McDonald's, 42, 174–75
McDonnell Douglas, 20, 30–31; Automation Company, 30
McGarry, Ray, 110–11
McGregor, Douglas, 65
McKinsey & Co., 61
McNamara, Robert, 145
Meadows, Edward, 33–43
Menk, Carl, 54
Mercedes Benz, 154, 158
Mesdag, Lisa Miller, 44–55, *49–50*
Meyer, Ira, 175
Meyers, Maurice, 131
Miami, Fla., 40
middle managers, 68
Milholland, Dick E., 48
Miller, Edward, 16
Mills, Ted, *63*, 67–68, 126
M.I.T. (Massachusetts Institute of Technology), 3, 16, 23, 65, 185; Sloan School of Management, 55, 84
Mitsubishi Motors, 95, 99
Mogensen, Allan H., viii–ix
Moline, Ill., 52
Monsanto, 139
Moore, Alicia Hills, 19–32
Moscone, George, 150
Moser, Bev T., 109
Moss, Stephen, 150
MOST (Columbus, Ohio), 129, 131
Motorola, 158–59
Murphy, James J., Jr., 129
Murrin, Thomas J., 72–73, 75, 79, 80

National Academy of Engineering, 95
National Academy of Sciences, 3–4
National Electrical Contractors Association, 132
National Federation of Federal Employees, 145
National Highway Traffic Safety Administration, 155
National Park Service, 145
national productivity program, 17
Navy, U.S., 144–45

NC machines. *See* numerically controlled machines
NCR, 126
nepotism, 120
Neukranz, Donald W., 71
Nevin, Joseph P., 82, 84, 85, 86–87, 91
New Providence, N.J., 85
New York Stock Exchange, 177
New York Telephone Company, 170
Nicol, Donna, 41
Nilson, Edwin N., 22
Nissan, 99, 100, 102
Nixon, President Richard M., 11
Norsworthy, J. R., 3
Northrop Corporation, 20; Aircraft division, 109
Northwest Traffic Associates, Inc., 87–88
Nothenberg, Rudy, 150
numerically controlled (NC) machines, 22, 23

Occupational Safety and Health Administration. *See* OSHA
office procedures reform, 81–91
Oil, Chemical and Atomic Workers, 131
oil price fluctuation, 13; and capital investment, 7
OPEC, 2
"open systems management," 58
Orlando, Fla., 173
OSHA (Occupational Safety and Health Administration), 9, 184; carcinogen standards, 9
Ouichi, William G., 58, 70, 73–75, 78, 80
output-per-worker rates, 2, *4, 5*

Pacific Telephone Company, 126
participative management, viii, x, 58, 59, *62–63,* 71, 74. 80, 113, 116, 121
Pascale, Richard T., 58
Peirce, James M., 145
Penney, J.C., 160
Pennsylvania, 161
Pentagon, the, 145
PEP (Beaumont, Tex.), 129, 131
Personnel Management, U.S. Office of, 140
Peters, Thomas J., 61
Petty, Ronald, 42
Phoenix, Ariz., 141, 148, 149, 151
Pillsbury, 42
Polaroid, 108
Pontiac, 57
Poplar Street Bridge (St. Louis, Mo.), 132
Power, J.D., & Associates, 154
Power Systems Company (Westinghouse), 73
Pratt & Whitney, 22
PRIDE (St. Louis, Mo.), 129–32, 135–39
Princeton University, 1
process planning, 24, 25, 28
Procter & Gamble, x, 66, 108, 139
productivity: and affirmative action, 13; and attitudes toward work, 11, 67; and bureaucracy, 81–91, *88–89*; and "buoyancy" factor, 10; and capital investment, 4, 5, 7, 17–18, 177–78, 181–82; and capital-labor ratio, 7, 17, 185; and cost-reduction programs, 34; and crime and vandalism, 12; cyclical influences on, 4–5; declining growth rate of, viii, 1–2, 3, *4, 5,* 10, 12–14, *15*; demographic influences on, 5–6; and diffusion of knowledge, 17–18; and employee-participation programs, 106–107; and equipment, 16; and factory management, 44–50; "gap," 43; in government sector, 140–51, *146*; and growth rate of economy (GNP), 9–10, 14, 177–82; increasing the growth rate of, 2–3, *4, 5,* 10, 13–14, 17–18, 32, 35, 50, 64, 79–80, 81–82, 106, *143, 187*; index of, 86–87; and inflation, 1–2, 7, 12, *15,* 34–35; institutions to enhance, 16; and intersectoral shifts, 8–9; and inventory, 97–98; and "just-in-time" system, 97–98; and manufacturing, 44–56; and nepotism, 120; and oil price fluctuations, 7–8, 13, *15*; and production managers, 54; and quality, relationship to, vii–viii, 51; and quality, theories on, ix; and regulations, federal, 9–10, 12–13, 14; and secretarial work, 78–79; and stagflation, 12; and tax changes, 14;and technology, 4, 16, 19–32, 35, 51, 65; white-collar, 36–37; and work schedules, 39; workers' attitudes toward, 67
Productivity Forum, the, 126–27

Project on Technology, Work, and Character (Washington, D.C.), 124
Public Systems Company (Westinghouse), 72, 73, 80

quality: Japanese standards of, 153, 163; problems in maintaining, 50; relationship to productivity, vii–viii, 51; and technology, 51
quality circles, x, 58–59, 74, 75–78, 107, 108–10, 118, 145, 159–60
quality control, 168–71
quality-of-worklife programs, 57–69, *62–63*, 103, 106, 111, 118, 119, 120, 123
quality standards, vii, 49, 51

Racine, Wis., 48
Raterink, Lee, 76
Rawlings, R. W., 99
Reagan, President Ronald R., x, 9*n*, 44, 128, 140, 141, 145, 146, 167, 170, 182–83
real output per worker-hour, 3
Regina floor-care appliances, 49, 50
regulations, federal, on business, 9–10, 12–13, 14
Reich, Robert B., 153, 192
Republic Steel, 54
research-and-development restaurants (Burger King), 41
research and development (R & D) tax credits, 186
Rice, Faye, 57–69, 106–17, 118–27
Riesman, David, 11
robotics, 51
Ronningen Research & Development Company, 30
Ronningen, Robert M., 30
Rosow, Jerome M., 126
Ross, Irwin, 128–39
Rudolph, John, 36, 37
Rukeyser, William S., vii–xi
Runyon, Marvin T., 99–100

Sachs, Jeffrey D., 188
St. Laurent, Norman, 115
St. Louis, Mo., x, 128–39
St. Louis Construction Users Council, 134
Salt Lake City, Utah, 23
Samuelson, Paul, 11
San Francisco, Calif., 149–50, 151
Santa Clara, Calif., 81, 90
Sauder, Gerald K., 129
Savas, E. S., 148
Scanlon, Frank, 173
Scanlon, Joe, 112
Scanlon Plan, the, 112–13
Scarsdale, N.Y., 67
Schlumberger, 30
Schmitt, Roland W., 23
Schroeder, Patricia, 141
Science Management Corporation, 7
secretary, role of the, 78–79
service sector, shift to, 8–9
"7-S" management model, 58
Shaughnessy, John W., Jr., 121, 126
Silber, John R., 149
Silverman, Elizabeth S., 128–39
"65 Billion Man-Hours" (Mogensen), viii
Skidmore, James, 7
Skinner, Wickham, 45, 46–47, 53
Smith, Donald, 40, 42
Smith, Peter, *49–50*
Solidarity (UAW newsletter), 125
Sony (San Diego, Calif.), 158
Sperry Univac, 32
Staats, Elmer B., 189–90
stagflation, 12
Starr, Martin K., 54–55
Stavrakas, E. James, 160
steel industry, 192
Stein, Herbert, 11
Steska, Edward L., 133
Strippoli, Gino T., 114–15, 116
Sturtevant division factory (Westinghouse), 78
Subaru, 154
Susman, Gerald I., 61
Sutherland, Ivan E., 23
Systems I to IV, 65

Tarrytown, N.Y., 112, 159
Tavistock Institute, 65
tax changes to promote investment, 14
Taylor, Frederick W., viii, 62–64, 66, 84, 159
Tektronix, 53
Telecommunications International Union, 118, 121, 126
Texas Instruments, 52, 193
Theory X and Theory Y, 65, 68
Theory Z, 70, 73
Theory Z (Ouichi), 73
"therblig," 63

Thompson, Debbie, 175
three-day work schedule, 39
Thurow, Lester, 8
Time Inc., x
Topeka, Kans., 107
Top Notch (Indianapolis, Ind.), 129, 131
Toxic Substance Control Act, 9
Toyota, 99, 102, 154
Trailways Bus Lines, 166–67
transportation industry, 164
Trist, Eric, *62–63*, 65
TRW, 53, 108, 158; Lawrence Cable division, 113–15
Tubbs, Robert J., 76
Tupper, Nathan G., 25
Turissini, Robert, 53
Twymon, Chuck, 112

UAW. *See* United Auto Workers
"under management," 36
Unigraphics system, 30
Unimation, Inc., 32
Union Jack (Denver, Colo.), 129, 130
United Airlines, 169
United Auto Workers (UAW), 57, 60, 103, 111, 118, 120, 121, 122, 123, 125, 159
United Food and Commercial Workers, 119
United Rubber Workers, 118; Local 87, 125
U.S.A., Inc., 191
U.S. Steel, 54, 134
United Steelworkers, 60, 118, 119, 123
University of California at Los Angeles, 73
University of Michigan, 65
University of North Carolina, 147
University of Rochester, viii, 29
University of Southern California, 84; Graduate School of Business Administration, 113
University of Tennessee, 156
utility industry, 164
Utterback, James, 16, 186

VeFac, 84
Vervinck, Bob, 111
Veterans Administration, 142
Vicksburg, Mich., 30
Volkswagen, 154, 157, 158, 161

Walker, Charles R. 64, 67
Wall Street Journal, The, 86
Walsh, Virgil, 136
Walt Disney Productions, 61
Walt Disney World, 173–74, 175
Warren, Alfred S., Jr., 103
Warren, Mich., 20
Washington Monument syndrome, 145
Washington University, 9
Watts, Glenn E., 124, 127
Wayne, Mich., 101
Weidenbaum, Murray, 9, 9*n,* 13
Western Electric, 53, 155; Hawthorne works, 64
West Germany, vii, ix, 29, 179, *180,* 186, 187, *187*
Westin Hotels (Western International holels), 176
Westinghouse Electric Corporation, ix, 59, 70–80
Wharton School, *63*
White, John P., 3
White, Robert B., 84–85
white-collar productivity, 36–37
Whopper Hotness program (Burger King), 33,41
Winpisinger, William, 119
Wisnosky, Dennis E., 25–26
Wofac Company, 84
word-processing center, 79
work, attitudes toward, 11
Work in America Institute (Scarsdale, N.Y.), 67, 126
"Working Smarter" series (Demaree, Burck, Main, et al), xi
"work innovation," 58
"work redesign," 58
work simplification, x
World Trade Center (New York, N.Y.), 57
World War II, 45
Wyoming, 189

Yale University, 64
Yankelovich, Skelly & White, 155
York University, *63*
Yoshida, Shuichi, 100
Young, John A., 44
Yugoslavia, *62–63*

"Z" management proposition, 58